Development of Environmental Policy in Japan and Asian Countries

Development of Environmental Policy in Japan and Asian Countries

Edited by

Tadayoshi Terao

and

Kenji Otsuka

palgrave macmillan   IDE-JETRO

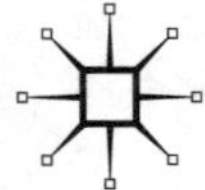

First published 2007 by
PALGRAVE MACMILLAN
Houndmills, Basingstoke, Hampshire RG21 6XS and
175 Fifth Avenue, New York, N.Y. 10010
Companies and representatives throughout the world

PALGRAVE MACMILLAN is the global academic imprint of the Palgrave Macmillan division of St. Martin's Press, LLC and of Palgrave Macmillan Ltd. Macmillan® is a registered trademark in the United States, United Kingdom and other countries. Palgrave is a registered trademark in the European Union and other countries.

ISBN-13: 978–0–230–00470–2
ISBN-10: 0–230–00470–9

This book is printed on paper suitable for recycling and made from fully managed and sustained forest sources.

A catalogue record for this book is available from the British Library.

Library of Congress Cataloging-in-Publication Data
Development of environmental policy in Japan and Asian
    countries/edited by Tadayoshi Terao and Kenji Otsuka
        p. cm.
    Includes bibliographical references and index.
    ISBN 0–230–00470–9
    1. Sustainable development—Japan. 2. Environmental policy—Japan. 3. Sustainable development—Asia. 4. Environmental policy—Asia. I. Terao, Tadayoshi, II. Otsuka, Kenji.
HC465.E5D48 2007
338.952′07—dc22   2006048298

10   9   8   7   6   5   4   3   2   1
16   15   14   13   12   11   10   09   08   07

Printed and bound in Great Britain by
Antony Rowe Ltd, Chippenham and Eastbourne

Contents

List of Tables

List of Figures

Notes on the Contributors

Yoshifumi Fujii Professor, Faculty of International Studies, Bunkyo University

Ryo Fujikura Professor, Faculty of Humanity and Environment, Hosei University

Nobuhiro Horii Research Fellow, Institute of Developing Economies, Japan External Trade Organization

Yasushi Ito Professor, Faculty of Commerce and Economics, Chiba University of Commerce

Michikazu Kojima Research Fellow, Institute of Developing Economies, Japan External Trade Organization

Kenji Otsuka Research Fellow, Institute of Developing Economies, Japan External Trade Organization

Hideaki Shiroyama Professor, Graduate School of Law and Politics, University of Tokyo

Tadayoshi Terao Research Fellow, Institute of Developing Economies, Japan External Trade Organization

Yuko Tsujita Research Fellow, Institute of Developing Economies, Japan External Trade Organization

Preface

This book is one of the outcomes of a five-year research project on environmental policy in Asia conducted at the Institute of Developing Economies, Japan External Trade Organization (Chiba) from the fiscal year of 2000 to 2004. Our research project focuses on the processes of environmental policy formation and implementation and their links to economic development, the objective of which is to reveal factors of dynamism between environmental policy and social change, in domestic, regional and global contexts, which could mutually influence effectiveness in environmental policy.

We have already published two volumes of Japanese books as IDE Research Series No. 527 in 2002 and No. 541 in 2005. Each paper in this book is revised and rewritten by each project member in English, based on his/her own Japanese article published in these books. In addition to these Japanese books, we have published an English report on *Studies on Environmental Pollution Disputes in East Asia: Cases From Mainland China and Taiwan* as Joint Research Program Series No. 128, in 2001, by our institute.

Among many books on environmental policy published in English, this book is unique in focusing on both present issues, which Asian countries are facing, and past experiences of Japan, which is a front runner in economic development in Asia, struggling against heavy environmental pollution under rapid economic growth after the Second World War. We believe this book could help those who are involved or interested in environmental policy reform to find any clue to promote environmentally sustainable development in Asia and other developing countries.

We would like to thank not only all our project members who have devoted themselves to dozens of project meetings for five years and paper writing under hard time constraints, but also numerous persons who have given us kindest support and advice on our research projects. We also are grateful to staff at SUNT Co. Ltd, PALGRAVE MACMILLAN and IDE-JETRO for generous support in editing this book.

TADAYOSHI TERAO
KENJI OTSUKA

Introduction

Tadayoshi Terao and Kenji Otsuka

In order to conquer poverty, raise social welfare and secure political stability, most governments in developing countries have promoted economic development through industrialization. In the process whereby developing countries achieve economic development through industrialization, environmental problems arise in various forms and the government, companies and citizens in each country are compelled to find countermeasures. Responses to environmental problems often involve trade-offs with the policy that promotes economic development. Fundamentally, economic development policies are, in many cases, major factors in the cause of environmental problems.

As economies become increasingly globalized, environmental problems are occurring across borders in conjunction with international trade and direct investment, and problems that cannot be tackled by a single country are increasing. Global environmental problems, such as climate change caused by greenhouse gases and destruction of ozone layers induced by chlorofluorocarbons, require greater participation by developing countries to develop remedies. However, the existence of domestic environmental policies within countries is a precondition for international measures, and international measures will not make progress unless domestic measures are conducted well by individual countries. Moreover, the conditions that must be taken into account when considering environmental policy at the national level are influenced by changes in international society.

In looking at environmental policy, we mainly focus on measures to reduce environmental pollution caused by economic activities. The gravity of environmental problems became widely known to the world through international conferences and environmental organizations in the early 1970s, and laws for pollution control were enacted not only in developed countries but also in most developing countries, with a focus on industrial pollution, including air pollution, water pollution, waste and noise. However, in many cases the government agencies that administer the legal system and the detailed legal provisions for enforcing the regulations did not receive the necessary enhancement. Except for a few successful cases,[1] most countries are still enduring poor implementation and enforcement of environmental law and policy, and the reasons for this are considered to be imperfect legal systems, lack of personnel, insufficient equipment, low budgets, economic development-oriented political forces and deficiencies in environmental awareness among entrepreneurs and citizens, etc.

In studying the environmental policy problems in Asia, this book focuses on the processes of environmental policy formation and implementation and their links to economic development. The objective is to reveal factors of dynamism between environmental policy and social change, in domestic, regional and global contexts, which could mutually influence effectiveness in environmental policy. Unlike many studies on environmental policy, we do not use any single existing discipline, but rather use a politico-economic approach to environmental policy processes related to economic development. In the context of this book, the politico-economic approach does not merely mean political analysis of economic policy, but means analysis of mutual interactions among a variety of stakeholders in the policy process, based on the structure woven by the incentives of each stakeholder.

In the case studies in this book, we focus on the two points below.

The first point on which we focus in the case studies is the experiences of industrial pollution control in Japan as an Asian country that realized rapid economic growth through the 1960s and into the early 1970s.[2] Although the Japanese experience with industrial pollution control has been mentioned frequently as an Asian model for environmental policy due to the similarities in rapid economic growth and industrialization in Japan and the current developing countries, we need to rethink the Japanese experience from the viewpoint of political economy to reveal how some measures have been used but others have not and how some measures have succeeded but others have not. In other words, we try to reveal both universal and unique factors in the Japanese experience in order to serve as an operational reference for developing countries in Asia and other areas.

The second focal point in our case studies is the problems of implementation and enforcement in environmental policy, especially in pollution control, in Asian countries. Although most Asian countries have set up environmental administration and environmental legal systems to deal with the deteriorating quality of the environmental in each country, most of them have not yet succeeded in controlling environmental pollution and destruction due to poor implementation and enforcement. To study this issue, it is necessary to investigate not only the current system of policy, law and administration but also to investigate the dynamic process of environmental policy formation and implementation, as well as their links to the broader range of social change, such as industrialization, democratization, decentralization and globalization, in each country.

This book consists of two parts. In Part I, we will rethink Japanese experiences in environmental pollution control mainly during the era of rapid economic growth after the Second World War. In this part, we will reveal factors behind the so-called successful Japanese experience with pollution control to share with other Asian countries, which are still tackling environmental pollution. In Part II, we will analyse the dynamism of the environmental policy process in Asia. In this part, we will examine how each factor, some of which are common in cases in Japan, exerts influence on other factors to lead environmental policies to success or failure in each country.

The outline of each chapter is presented below.

Chapter 1 attempts to examine the so-called 'Japanese experience' of industrial pollution control and attempts to position it in the process of rapid economic development of late-comer industrialization. From such a point of view, the industrial policy that promoted rapid industrialization becomes an important factor in understanding industrial pollution control policy in Japan. Industrial pollution control policy in Japan was formed and promoted as a part of industrial development policy, using the same policy tools as industrial development policy, such as low-interest loan programmes and preferential tax treatments.

Chapter 2 examines the historical dynamic process of the end-of-pipe technology development under the command and control (CAC) approach in Japan, focusing on SOx and NOx control. Modern economic theory tells us that economic instruments would be more efficient than CAC in environmental policy because CAC would bring no more incentive than regulation. However, this chapter carefully examines, through case studies of air pollution control in Japan, how the CAC approach has brought about development of pollution control technology.

Chapter 3 focuses on the role of local governmental research institutes in technology development and diffusion for pollution control in Japan. Local governmental research institutes (LGRIs) were established mainly in each prefecture and some large cities, in order to develop the technological state of regional industries by conducting technical guidance, tests on request and research and development (R&D). This chapter describes how important the role of LGRIs has been in Japan, especially for small and medium enterprises, and examines the implications of this for developing countries.

Chapter 4 examines the effectiveness of administrative guidance issued by local governments for air pollution control to industries in Japan. Although administrative guidance in environmental policy is often seen as inefficient both economically and environmentally, it played the central role in industrial pollution control during the era of no national regulation in some cities such as Osaka and Kitakyushu. Through case studies in these two cities, this chapter discusses the factors that have led to the success of administrative guidance, as well as the effectiveness of and implications for technical cooperation with developing countries.

Chapter 5, as the first chapter of Part II of this book, examines the problems in the existing air pollution control in China and discusses the possibility of installing economic instruments while considering the characteristics of China's transitional economy in the reform era. The case study of air pollution control in China presented here suggests that China would better realize medium to long-term policy effectiveness by adapting some economic incentives together with certain institutional reforms, rather than enhancing command and control policy.

Chapter 6 focuses on a rating programme for industrial pollution control based on information disclosure in Indonesia. PROPER (Programme for Pollution Control, Evaluation and Rating), initiated in Indonesia and supported by the World Bank, has attracted international attention and is now carried out in the Philippines, India and China, too, but its historical background and policy development in Indonesia have not been clarified. This chapter re-examines rating programmes in different stages, including PROPER, and discusses their meaning and limits for environmental management in Indonesia.

Chapter 7 examines the role of public interest litigation in industrial pollution control. In India recently, many environmental lawsuits are heard by the Supreme Court every Friday, now known as 'Green Friday', and most of them are brought as public interest litigation (PIL), which allows the general public to invoke the warrant jurisdiction of the court. Although it can be said that PIL plays a significant role in raising people's awareness of environmental issues, it should be noted that the judgements are not always implemented well. This chapter also gives attention, through case studies, to the role of local communities as informal regulators.

Chapter 8 reviews the transformation of governance systems in environmental policy implementation, focusing in particular on industrial pollution control and examines the effectiveness and tasks of multi-stakeholder governance in China. Since the 1990s, the central government has enhanced regulatory enforcement for industrial pollution, while reforming the governance system of environmental policy implementation with the involvement of the People's Congress, mass media and NGOs. Following a macro review of such transformation, this chapter discusses its effectiveness from the viewpoints of regulatory enforcement, incentives for firms and public participation.

Chapter 9 attempts to present alternative views of environmental policy formation in the process of democratization from an authoritarian regime in Taiwan and also examines the major achievements and problems of environmental policy in Taiwan following completion of the democratization process there. In Taiwan, from the first half of the 1990s, political liberalization and democratization helped development of environmental policy administration and institutions. However, after the late 1990s, side effects of rapid democratization became increasingly prominent in many aspects of society, including local environmental policies, which were diversified significantly after decentralization.

Chapter 10 analyses the characteristics of the environmental cooperation regime of the East Asia region, in comparison with the environmental cooperation regime of Europe, and also analyses the effectiveness of Japan's policy towards regional environmental cooperation in East Asia, mainly through studying a case in China. This chapter examines not only international and regional regimes in environmental policy but also domestic institutional factors in policy implementation in China.

These ten chapters suggest a number of points, including those below.

The first point relates to the possibilities and limitations of applying Japanese experiences to pollution control policies as a model for environmental policy during rapid industrialization. It is, indeed, possible that Japan's experiences discussed in Part I could provide an interesting viewpoint on the effectiveness of pollution control based on industrial policy by administrative dictate. However, Japan's experiences can also reveal limitations in administration-dictated industrial policy that hamper institutional reforms for broader stakeholder participation and decentralized decision-making in environmental policy formation, including policies on environmental impact assessment, sustainable urban planning and so on.

The second point, as discussed in Part II, is that the socio-economic conditions behind Japan's 'successful' pollution control policy will not necessarily be identical to those in developing countries in Asia. First, regulatory enforcement is not perfect in many countries, due to lack of governmental resources, political will and social pressure. Second, many countries are already shifting their environmental policy from CAC to a policy mix with market-based instruments (MBIs) and information disclosure. Third, although many countries are basing their environmental policy on market-based instruments (MBIs) and multi-stakeholder governance with information disclosure and public participation, the effectiveness of such new environmental policy instruments is not adequate due to imperfect political and economic reform in each country.

The last point, as discussed in the last chapter of this book, is that the effectiveness of environmental cooperation in East Asia is constrained by such factors as weak ad hoc regimes and scant interactions with comprehensive regimes like APEC. Focusing in particular on Japanese environmental aid to China, which is deeply embedded in environmental cooperation in East Asia, it is suggested that consideration should be given both to domestic factors in China (such as environmental regulations, financial environment, organizational coordination, international rules and roles of NGOs and foreign companies), as well as to strategy for regional cooperation by Japan. Japan's strategy is often to promote the Japan model, which developed under its own particular conditions, with its strict command and control, best available 'expensive' technology and so on, but such a strategy seems to constrain the effectiveness of environmental aid.

In sum, in order to resolve the environmental policy problems in Asia, we should rethink Japanese experiences in environmental pollution control and also carefully analyse socio-economic factors in the environmental policy process in each country. We will be deeply gratified if we can provide some insights for further discussion on improvement of the current environmental situation in East Asia through this book.

Notes

1. Singapore is often said to be one of the successful countries in implementing effective environmental management through government initiative. See Owada (1993) and Heng (1997).
2. For example, see Kojima *et al.* (1995).

References

Heng, Lye Lin (1997) 'The Enforcement of Environmental Law in Singapore', in Yoshihiro Nomura and Naoyuki Sakumoto (eds), *Environmental Law and Policy in Asia: Issues of Enforcement*, Tokyo, Institute of Developing Economies.

Kojima, Reeitsu, Yoshihiro Nomura, Shigeaki Fujisaki and Naoyuki Sakumoto (1995) *Development and the Environment: The Experiences of Japan and Industrializing Asia*, Tokyo, Institute of Developing Economies.

Nomura, Yoshihiro and Naoyuki Sakumoto (1997) *Environmental Law and Policy in Asia: Issues of Enforcement*, Tokyo, Institute of Developing Economies.

Owada, Takiyoshi (1993) *Eko deiberoppumento – singaporu, tsuyoi seifu no kankyou jikken* (Eco-development in Singapore: an environmental experiment by a strong government), Tokyo, Chuokoron-sha.

Part I

Rethinking of Japanese Experiences in Environmental Pollution Control

1

Industrial Policy, Industrial Development and Pollution Control in Post-war Japan: Implications for Developing Countries

Tadayoshi Terao

Introduction

There is a self-serving belief that the 'Japanese experience' with industrial pollution was a paragon of success in conquering environmental problems encountered in the process of rapid economic development and that this experience can be transplanted to late-comer developing countries. On the other hand, there is also the idea that the 'Japanese experience' should be presented to the developing countries as a negative lesson on serious pollution damage and failure of environmental policy. It is possible to view both of the arguments as standing on the common premise that Japan's experience can provide some kind of model for environmental pollution and policy.

However, we first have to consider why the Japanese experience can be generalized as a 'model'. Before we attempt to consider the Japanese experience as a model, the generation of industrial pollution and the process of implementing countermeasures must be analysed and appropriately positioned in the chronological process of economic development from the rapid-growth era or even earlier. Otherwise, the Japanese experience is merely a story of anecdotal experiences.

Below, we will examine whether a certain 'model' can be constructed by positioning the Japanese experience as a process through which a late-comer country aims at industrialization, and by characterizing industrial pollution as a problem generated in the process.

Generally, it is possible for a late-comer country, through industrial policy, to utilize the experience of advanced countries and to transfer technology that has already been developed to achieve industrialization more quickly.

Since the late-comer country can predict which industries can enjoy the benefits, such as decreasing costs over a long term, of introduction of technological knowledge from an advanced nation, the late-comer country can pursue economic growth systematically by promoting such industries using industrial policy measures. As industries that can be expected to enjoy decreasing costs continuously over a long term, the heavy and chemical industries are preponderantly chosen as the industries to be promoted. The heavy and chemical industries are potential pollution sources, and if neither environmental impact assessments nor measures against pollution are introduced appropriately, they may cause pollution damage to the surrounding environment and residents. If an industrial policy to promote the heavy and chemical industries is carried out without fully taking the industrial pollution accompanying it into consideration, as seen in Japan in its high-growth era, the success of the industrial policy may cause aggravation of industrial pollution.

On the other hand in Japan, in order to compensate for the delayed response to industrial pollution, private enterprises were obliged to make large-scale investments in the mid-1970s. On that occasion, the government supported the pollution prevention efforts of private companies through measures such as administrative guidance, low-interest loans and preferential tax treatment, etc. In Japan, it can be said that industrial pollution regulation was incorporated into industrial policy and was promoted as a part of it.

In this chapter we will examine the relationship between industrial policy and industrial pollution in Japan, from the viewpoint of 'development and environment'. In Section 1, we will show how industrial policy played an essential role in the Japanese experience of industrial development by introducing the concept of 'developmentalism'. In Section 2, the relationship between industrial policy and industrial pollution will be analysed. It is assumed that industrial policy should affect the pattern of industrial development and industrial structure. The effect of the Japanese government's emphasis on heavy and chemical industrialization in its industrial policy during the rapid growth period will be discussed. Next, we will shift our focus to the unique characteristics of 'industrial pollution regulation as a part of industrial policy' in Japan in 1970s. In Section 3, energy policy, as an important element of both industrial development policy and pollution control policy, will be discussed. We will show how the complex interests of the energy industry affect the direction of air pollution control policy, especially in case of SOx emission regulation. In Section 4, we will show how industrial pollution could be incorporated into the concept of 'developmentalism' and discuss the relationship between industrial policy and industrial pollution control.

1. 'Developmentalism' as a system and industrial policy

The experience of economic development in post-war Japan can be considered to have been a process of catching up with the advanced countries through a series of policies based on the system of 'developmentalism', a system that intensively pursues industrialization. If the pattern of development pursued by 'developmentalism' affects the process of industrial pollution generation and the possible measures taken against it, then we could stress the importance of a 'Japanese experience' of industrial pollution control, in contrast to other developed countries.

First, the possibility of economic development pursued by the policy system based on 'developmentalism' will be discussed below.

In many types of manufacturing industries, a continuous decrease in costs over the long run can be anticipated by successively introducing advanced technologies, which were developed and put to practical use over many generations in developed counties. Therefore, it is relatively easy for policy-makers in late-comer countries to predict for which industries the long-term marginal (and average) costs will diminish by observing the experience of advanced countries. In such cases, it is possible to provide the fundamental conditions for introducing 'developmentalism' as a system; namely, a series of policies ultimately aimed at forming a national economy by strategically fostering promising industries through industrial policy, intensively providing domestic physical infrastructure, and improving the educational system to cultivate human capital. The policy system based on 'developmentalism' consists of government intervention in the market economy, under the condition that private property and market competition should be maintained as part of the fundamental framework, and this policy system is set up with the aim of achieving industrialization that can be measured objectively as sustainable growth of per capita income. It is impossible to implement such policies without minimum political integration of a state.[1]

The components of the model of 'developmentalism' are as follows.

Narrowly Defined Industrial Policy Components
1. The principle of market competition based on the system of private property rights;
2. Enforcement of an industrial policy by selecting industries where costs will decrease over the long term;
3. Exports to achieve the goal of industry promotion;
4. Promotion of small and medium-scale enterprises.

Distribution Policy Components
5. Raising of domestic demand through consumption by equalizing income distribution;
6. Equalization of the agricultural land distribution.

Social Infrastructure Components
7. Enhancement of the educational system;
8. Establishment of a fair and modern bureaucracy system that abolishes nepotism.

Industrial policy should be an essential element of 'developmentalism'. However, we should not forget that 'developmentalism' emphasizes the resource allocation function of the market mechanism, and an industrial policy should always be an intervention conditioned on the existence of the market mechanism. 'Developmentalism' is reinforced and kept sustainable as a development policy through interaction among the various elements mentioned above. When selective industrial policy intervention succeeds, it will cause unequal income distribution, which could bring on social destabilization. An income distribution policy is necessary to maintain the series of policy systems of 'developmentalism', by alleviating the negative impact of rapid economic growth.

On the other hand, when general demand for consumption is created and the equalization of the income distribution could create a consumption demand among the general public, then that may become the base of the industry development. The industrial policy and the distribution policy are reinforced mutually by the interaction between them. Provision of a social infrastructure, such as enhancement of the education system and development of a trustworthy and capable bureaucracy, is a precondition for the success of industrialization through industrial policy. Social infrastructure is also important for implementing a distribution policy and, reciprocally, equalization of income distribution expands opportunities for education, which helps provide a supply of human capital to enterprises and the bureaucracy.

Yonosuke Hara modelled the economic development of East Asia (Japan, South Korea and Taiwan) as follows, in support of an argument for 'developmentalism':

> It can be said that the market economy system that is peculiar to the countries in East Asia is an economic system which realizes efficient use of mass-production-type transferred technologies from advanced nations and consequently is designed to accelerate economic growth. In a situation where only advanced nations possess technological knowledge in the international economy, the promotion of domestic industries in East Asian countries became very difficult due to the free market mechanism. In such cases, positive policy intervention by the governments of East Asian countries has been carried out. Industrial organizations have been formed with the participation of corporate organizations with strong internal cooperative relations and sub-groups of those organizations that complement each other in terms of the government's policy intervention. Such industrial organizations are also compatible with the economic and social motivation of many individuals.[2]

'Developmentalism' is a system of policies for mobilizing resources towards the aim of economic growth, and for maintaining rapid growth. Usage of the experience of economic development in Japan was not necessarily the only strategy used by latecomer countries for industrialization. It will be exceptional to know exactly the path to industrialization by 'the standard tactics which latecomer countries should step on', i.e., industrial policy, and to have done it in the form as extremely put into practice.

2.   Industrial policy and industrial pollution

In this section, the influence of industrial policy on industrial pollution is considered.

Since industrial policy is only a means for promoting achievement of 'desirable' industrial structure, even if it were not for industrial policy, there might be no major change in industrial structure in the long run.

Industrial policy may influence industrial pollution indirectly by changing industrial structure and industrial organization and then, finally, affecting the state of industrial pollution through those changes.

Moreover, in Japan, after the government recognized the seriousness of industrial pollution, the industrial pollution regulation policy was promoted as a part of industrial policy. We will examine the validity of the industrial pollution regulation by looking at industrial policies, such as administrative guidance, low-interest loan programmes and preferential tax treatments.

Industrial policy and heavy and chemical industrialization

Heavy and chemical industrialization as a policy aim

Implementation of industrial policy is the most important constituent factor of 'developmentalism' as a system. According to microeconomic theory, the industries which should be preponderantly promoted by industrial policy are those in which decreasing long-term costs can be expected. In the early stage of industrialization, which is where industrializing late-comer countries are, decreasing long-term costs can be expected in many manufacturing industries, especially in labour-intensive light industries. However, it is most likely that the heavy and chemical industries may be counted upon to enjoy decreasing long-term costs continuously, after the late-comer countries have enjoyed a comparative advantage in labour-intensive light industries.

In the experience of industrial policy in Japan, in the 1960s when heavy and chemical industrialization was promoted, the 'income elasticity standard' and the 'rate-of-productivity-growth standard' were well known as standards of industrial selection. Moreover, in the report entitled 'The Industrial and International Trade Policy of the 1970s', which set a vision for the 1970s, the Industrial Structure Council announced, in the last stage of the high-growth era, measures for 'excessive concentration and environmental standards' (i.e., reduction of congestion in urban areas, reduction of pollution problems and increased efficiency in use of energy resources) and the 'labor contents standard' (i.e., the provision of good workplace environments).

However, if the income elasticity of a product and the rate of productivity growth of an industry are high enough, such an industry should grow by itself, and there is no theoretical reason for having to treat such an industry favourably in political terms.[3]

It was thought that the heavy and chemical industries were industries that met both standards.[4]

Many companies in the heavy and chemical industries are types of companies that easily generate industrial pollution, and if heavy and chemical industrialization is promoted too quickly and too extensively without taking the proper measures against industrial pollution, it is clear from the experience of advanced countries that serious industrial pollution will be generated. However, until the middle of the 1960s, Japan was promoting heavy and chemical industrialization. People who grasped the situation correctly were not in the majority, and the necessity for industrial pollution measures was not widely recognized. Huge growth in production facilities and technology was pursued in the heavy and chemical industries, which are industries that typically display decreasing long-term costs.

Plant and equipment investment adjustment and over-investment

In raw material industries that are processing industries with economies of scale, such as oil refining, petrochemicals, iron and steel, and synthetic fibre, the industrial policy authorities emphasized the necessity for 'plant and equipment investment adjustment'.

Plant and equipment investment adjustment was called 'adjustment within industry' in accordance with production control and price adjustment, and was considered to be a typical form of industrial policy.[5] Below, we will examine the effect of the plant and equipment investment adjustment on the industrial development of heavy and chemical industries.

When the Japanese industrial policy authorities attempt to justify market intervention, 'excessive competition' has always been asserted. Excessive competition is also considered to be a major factor in excess capacity, dumping, destructive price competition and low profits of companies.[6] However, the policy authorities did not necessarily define 'excessive competition' clearly. While the cause of excessive competition was an overly small scale, it was presupposed, conversely, that excessive competition was the cause of the small scale.

Excessive competition was explained using the decreasing cost hypothesis by Yasusuke Murakami. Market share maximization behaviour and profits maximization behaviour are not distinguished in long-run average (and marginal) cost decreasing industries in microeconomic theory. The method of increasing profits is to expand market share. When the decreasing cost is known over the long term, a market price cannot maintain static equilibrium but becomes unstable, and competition should be intensified.

The measures that the policy authorities, the Ministry of International Trade and Industry (MITI), actually took in order to prevent excessive competition consisted of creation of an anti-recession cartel and 'plant and equipment investment adjustment'. Theoretically, excessive competition will occur in the market with an oligopoly, such as raw material industries that produce homogeneous products using a large fixed capital. In that case, the policy authorities may control 'excessive competition' by market intervention, which is a kind of complementary industrial policy, and may raise social welfare as a result. However, in order to determine the socially optimal level of fixed capital investment as a whole and the number of entry companies, the policy authorities have to possess detailed knowledge of the technologies and detailed information on the management of individual firms. This is impossible in reality.

The method wherein the policy authorities prepare opportunities for each industry to implement 'voluntary adjustment' of plant and equipment investment tends to reduce the cost of information-gathering compared to the policy wherein authorities allocate each enterprise's share of plant and equipment investment directly. However, during the process, each enterprise could gather the information on other's investment programmes and could share the arguments on adjustment of plant and equipment investment among the industrial community, so that the process would be a precise process for removing the information uncertainty regarding investment that individual enterprises face. The uncertainty regarding plant and equipment investment leads to wide variation in the market conditions that are anticipated by each enterprise and may cause a natural 'time lag' in investment. If an adjustment opportunity finishes with mere information exchange and without voluntary adjustment being successful, it only decreases uncertainty, erases the natural 'time lag' of investment of each enterprise, and has the effect of urging them to invest all at once.[7]

In both the case of adjustment by the policy authorities and the case of voluntary adjustment by enterprises, the 'share principle', which allocates the amount of investment permissions based on present market share of an established company, was often used upon actual approval and a quota rule. The consequence of this is paradoxical. Each enterprise desires to have a production capacity exceeding optimum levels to expand market share, in order to receive a more advantageous investment permission allocation. Therefore, excessive competition of a much more serious nature occurs as a result.

It is thought that such a mechanism was generated in iron and steel, petrochemicals, oil refinery and other industries, which were the raw material industries where plant and equipment investment adjustment was attempted by a policy to control excessive competition. Plant and equipment investment adjustment and over-investment in the petrochemicals industry in the high-growth era are described below, as an example.[8] The characteristic feature of the petrochemicals industry in the high-growth era was rapid growth with active new entries launched by domestic business groups.

The 'Foreign Investment Law' was the basis of the authority for the intervention by the government in the petrochemicals industry, which depended heavily on imported foreign technologies. The Foreign Investment Council had opted for foreign technology importation before capital liberalization began in Japan. When a company attempted to join the industry as new entrant, it had to obtain governmental permission to establish a joint corporation with foreign capital and to invest in plant and equipment.

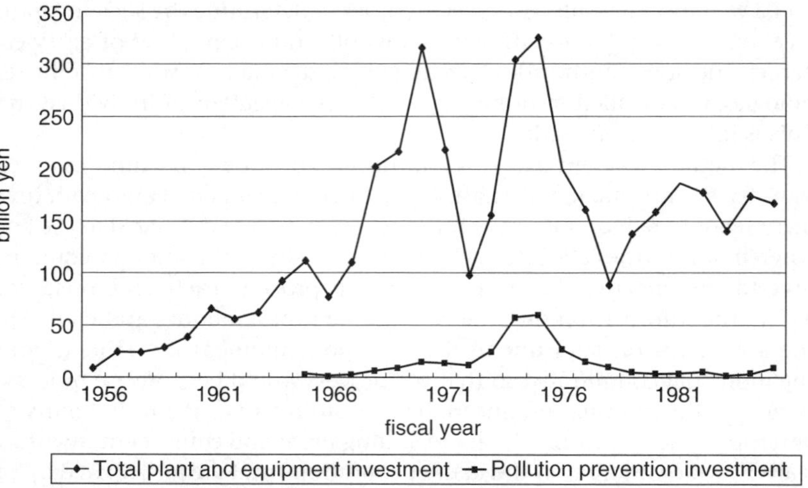

Figure 1.1   Total capital investment and pollution prevention investment of petro-chemical industry
Source: Ministry of International Trade and Industry, *Shuyo Sangyo no Setsubi Toshi Keikaku*, various issues.

In the case of the petrochemicals industry of the high-growth era, the government-business complex informal roundtable group made its fixed capital invest all at once, thereby destroying the natural 'time log' in plant and equipment investment that normally results from the different predictions of each enterprise due to information uncertainty. Moreover, the minimum capacity standard for new entry set by the government eliminated the difference in decisions on investment scale among the enterprises and concentrated investments at a similar scale (around the minimum standard). The companies adopted the strategy of expanding their market share to secure future plant and equipment investment permission allocation. All of those factors brought about over-investment. Moreover, the minimum capacity standard, which was expected to act as a barrier to excessive entry, was overcome contrary to MITI's expectations by almost all the enterprises that hoped to enter. To achieve this, companies pursued the corporate strategy of 'the full set principle', where each business group provides the necessary financing and simultaneously aims at market share expansion (Figures 1.1 and 1.2).

It is thought that the above-mentioned excess investments and destruction of natural investment rhythm were generated through an almost similar process in plant and equipment investment adjustment in the iron and steel industry too.[9]

It is possible that industrial development through capital investment competition under severe financial restrictions made it difficult to sufficiently

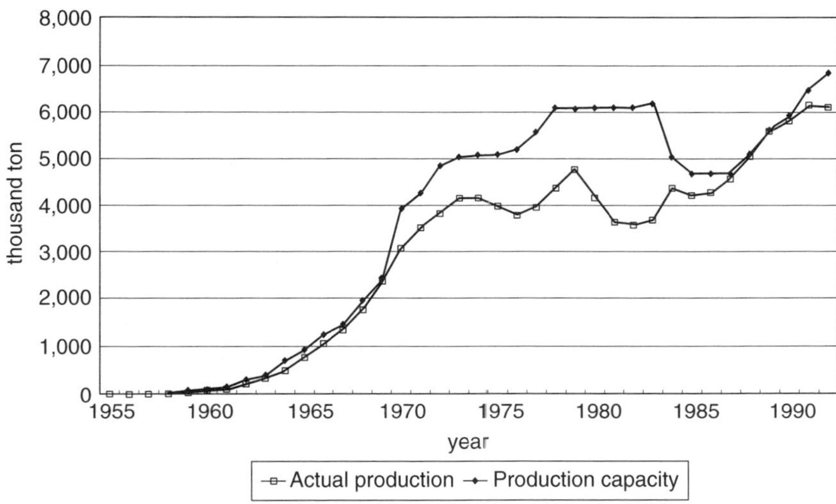

Figure 1.2   Production capacity and actual production of ethylene
Sources: Japan Petrochemicals Industry Association, *Sekiyu Kagaku Kogyo*, 30nen no Ayumi, 1989;
Ministry of International Trade and Industry, *Sekiyu Kagaku Tokei Nenpo*, various issues.

invest in pollution control facilities in the cases of the petrochemicals and
iron and steel industries and ultimately enlarged the load on the environment. Heavy and chemical industries in particular are accompanied by the
large-scale regional development projects that include seashore reclamation
in many cases. Most likely, the capital investment competition was accompanied by a rapid expansion in scale that brought about superfluous regional
development, without the preliminary environmental impact assessments
being put into practice.

Industrial pollution regulation as a part of industrial policy

Private enterprise was forced to make large-scale investments for industrial
pollution prevention measures in the period when rapid growth of the
Japanese economy began to slow. Private enterprise made use of the various
industrial policy measures and the experience accumulated in industrial
development policy prior to that. In Japan, it can be said that industrial pollution regulation was incorporated in the system of industrial policies and
was promoted as a part of that policy system. Given that industrial pollution
regulation policy was delayed at that time, 'industrial pollution regulation
promoted as a part of industrial policy' was an effective method, at least over
the short term.

If the industrial pollution regulation policy is incorporated as a part of
industrial policy, then how is it positioned in the framework of the theory of
industrial policy? First, although industrial policy is divided into 'strategic

industry policy' and 'complementary industrial policy', industrial pollution regulation is a policy that prevents negative 'externality' and is clearly classified as 'complementary industrial policy'.[10]

For example, when industrial pollution regulation is implemented among industries that are a significant pollution source, there may be expected to be some industries whose comparative advantage domestically might decline and be lost. For such an industry, a structural adjustment programme will also need to be implemented as a 'complementary industrial policy'. In an area with a high population density, as in Japan, the cost of pollution prevention is higher than in an area of low population density, and the comparative advantage of the pollution-intensive industries will be lost easily by introduction of pollution regulation. Most likely, the long-term strategy of the enterprises belonging to such a pollution-intensive industry will be relocation to overseas locations with more advantageous production conditions, including regulation of pollution emissions, or withdrawal from the industry. Of course, the cost of pollution prevention is only one factor that determines the comparative advantage of an industry.

As most industrial pollution regulation has a clear legal basis, pollution regulation may be considered a 'mandatory-type policy', although industrial policy is divided into 'discretionary type' and 'mandatory type' by the policy authorities. However, discretionary-type policies, which the legal system copes with by issuing administrative guidance, were implemented when the government did not have sufficient authority to implement a mandatory-type policy. Moreover, even after the legal system was improved, the mandatory-type policy with administrative guidance was often implemented.

Furthermore, industrial policy is divided into that implemented by 'regulatory means' and that implemented by 'inductive means'. Industrial pollution regulation policy carried out through various kinds of economic preferential treatment is said to be implemented by inductive means, and that which utilizes direct command and control regulation is said to be implemented by regulatory means. However, preferential treatment was used as an inductive means, and policy implementation through collection of a pollution tax, i.e., indirect regulation, was not even attempted in Japan until the late 1990s. It is one of the features of industrial pollution regulation in Japan that the authorities have pursued a combination of direct regulation and preferential treatment to the present day. Conversion to indirect regulations, such as taxation of pollutant emissions and the set-up of a transferable emissions-rights market, has not been tried in Japan.

The various means of implementing 'industrial pollution regulation as a part of industrial policy' in Japan were direct 'command and control', including '*gyousei shidou* (administrative guidance)' to the industry and/or to each enterprise, and preferential treatment such as low-interest loans and special taxation measures. Preferential treatment complements direct regulation and is premised on the promotion of direct regulation, as far as that is possible. The preferential treatment and the direct regulation of the industrial pollution regulation policy were combined together in many cases, as a 'carrot-and-stick approach'.

Moreover, in conjunction with the promotion of pollution prevention investment by private enterprise, the development of the pollution prevention (industrial machine) industry and the pollution prevention engineering industry was indispensable. Various industrial policies for development of those industries were implemented, based on the Machine Industry Promotion Temporary Measures Law. The standardization of anti-pollution devices and measuring devices was carried out as well and contributed to the formation of those markets. Moreover, needless to say, the policy of promoting pollution prevention investment by enterprises stimulated the demand for the products of the pollution prevention industry and also contributed to the growth of those industries.

In the following section, we will show how the various means for implementing industrial policy could be utilized to implement industry pollution regulation policy in Japan.

Direct command and control with administrative guidance

According to the OECD's review of environmental policy in Japan published in 1977, on the basis of the direct regulation, 'the emission standards are enforced by persuasion rather than coercion'.[11] This is because the application of penalty is considered to be the last resort, to be used only when a polluter doesn't obey an order issued by the proper government agency for improvement or suspension of discharge. The enforcement of the law is conducted primarily through administrative guidance, rather than by punishing the polluter. The OECD report on environmental policy in Japan characterized direct intervention in private enterprise as a form of industry pollution regulation, as follows:

> Compliance and cooperation by industry cannot be entirely accounted for by good sentiments. It is also obtained by a delicate handling of sticks and carrots by the administration, which can be more or less responsive to industry's needs in terms of accelerated depreciations, of funding, of procurements, of permits, and the like. Pollution abatement efforts are part of a package negotiated between industry and administration. It has even been suggested that some segments of the Japanese administration welcomed pollution controls because it increased their bargaining power at a time when other types of controls, such as controls over imports, foreign exchange, licensing, had been relaxed.[12]

> Standards are utilised as a weapon in the hands of the administration in the negotiations it engages in with polluters rather than as prescriptions that automatically apply. There is no reason to believe that it is a less efficient way of utilising standards.[13]

Direct regulation using emission standards was implemented, combined with preferential treatment such as low-interest loans from government-affiliated financial institutions and special taxation measures. MITI received applications and played a gate-keeping role for access to low-interest loans from the Japan Development Bank (JDB), which provides the major portion of low-interest loans for large enterprises, and, indeed, MITI had a substantial role in deciding the allocation of the loans. Although most of the prefectural and municipal governments prepared individual low-interest loans too, they were for small and medium-sized enterprises and self-employed individuals with limited budgets. Furthermore, at that time, MITI was issuing guidance on pollution prevention to each industry and providing assistance for development of pollution prevention technology.

Starting in the early 1960s, MITI was in charge of air pollution prevention together with the Ministry of Health and Welfare, and MITI was in charge of water pollution prevention, in cooperation with the Economic Planning Agency. MITI lost most of its authority for direct regulation of industrial pollution due to legal revisions, pollution-related enactments in the 'Pollution Diet' in 1970, and the establishment of the Environment Agency in 1971. MITI's authority was shifted to the Environment Agency and local government. From that time on, pollution control measures by MITI tended to take the form of administrative guidance to the industries and private enterprises, through provision and allocation of low-interest loans and preferential tax treatment. It can be interpreted from the expansion of the low-interest loans and the preferential tax treatment in the first half of the 1970s that MITI was retrenching its operations as it had lost direct legal authority in industrial pollution control regulation. Although the need for these measures weakened in the latter half of the 1970s, MITI's earlier expansion of the preferential treatment made it difficult to reduce them.

The industrial pollution control measures for each enterprise can be carried out effectively only if the regulation is done comprehensively, so that it applies either on an industry-wide or area-wide basis. Individual private enterprises have a disincentive to spend large amounts on pollution prevention investment, given the severe competition with other enterprises, as it would handicap them in the competition. Individual enterprises will decide to invest in large-scale pollution prevention facilities only when all enterprises in the same industry are pressured to invest simultaneously.[14] Therefore, as far as the command and control approach by direct regulation is concerned, the industrial policy of MITI, which issued administrative guidance to each industry by establishing or utilizing industry organizations for every industry, can be thought to be an effective method for industry pollution regulation as well.

Environmental standards and emission standards in Japan are not uniform throughout the country. There are many variations; in some cases regulations are stricter and, in other cases, exceptions are made and regulations are partially relaxed (as was the case in the K-value regulation and the fuel low-sulphurization plan). These exceptions offer room for discretion to be more careful and effective during the process of implementing industrial pollution regulations through administrative guidance. The 'pollution control agreements' that local governments concluded with private enterprises may be considered as a kind of discretionary regulatory measure.

Preferential treatments

According to the 24th article of Basic Law for Environmental Pollution Control on the installation and maintenance of the pollution prevention facilities, 'administrative authorities need to make the necessary effort to take financial and taxation measures'. It is prescribed that the administration should 'provide the necessary funds and provide technical assistance, etc.', in the Air Pollution Control Law, which is a substantive law of the Basic Law for Environmental Pollution Control. There is a similar statement in the Water Pollution Control Law as well.[15]

As preferential treatment for pollution prevention investment by private enterprises, there are the low-interest loan programmes and the tax relief available through various types of preferential taxation measures. Low-interest loan programmes offered by a government-affiliated financial institution have an effect similar to a subsidy to private enterprises by reducing the interest payments. The interest rates of those loans are significantly lower than the market rate at the given time. Furthermore, in some cases these loans are interest free, such as loans to small and medium enterprises from local governments.

Although various measures, such as direct tax deductions, special (accelerated) depreciation and fund reserves, are combined, preferential tax measures basically reduce the tax payment burden of the private enterprises. The preferential tax measure system also has the same effect as provision of subsidies by substantially reducing part of the tax payment burden. Preferential tax measures are known as 'the tax expenditure'.

As for preferential treatment for environmental pollution prevention, direct subsidy to private enterprises from the central government has never been spent to promote investment for pollution prevention. However, for research and development on pollution prevention technology, substantial assistance was given by the central government to private enterprises. Local governments supplied subsidies mainly for the small and medium enterprises to prevent industrial pollution. In some cases, subsidies by local governments, combined with the provision of loans from them to small and medium enterprises for pollution prevention investment, were intended to reduce the amount of loan interest payments. In the latter half of the 1970s, the amount of subsidy payments by local governments for pollution prevention was about 10 billion Japanese yen annually.

We will present an outline of the low-interest loans and the preferential tax treatment for pollution prevention, as well as the effect of such treatment, in the following section.

Low-interest loan programmes.   It was in the fiscal year (FY)1960 that the low-interest loans offered by the Fiscal Investment and Loan Programme (FILP) started to be applied to pollution prevention investment of private enterprises. In the same year, Japan Development Bank (JDB) set up a loan scheme for investment in wastewater treatment facilities. Most of the loans offered by JDB are for large enterprises. The low-interest loans offered by the Pollution Control Service Corporation (later known as the Japan Environmental Corporation [JEC] and at present known as the Environmental Restoration and Conservation Agency [ERCA]) covered not only small and medium enterprises, but also large enterprises as well.

The loan schemes for pollution prevention provided exclusively to small and medium enterprises were set up in FY1960, through the Small and Medium Enterprise Modernization Promotion Fund. After that, the Japan Finance Corporation for Small Business (JFCS), Japan Small Business Corporation (JSBC) and Pollution Control Facility Lease Programme, etc., together with the Pollution Control Service Corporation, inaugurated funding to small and medium enterprises for pollution prevention.

The undertakings of the Pollution Control Service Corporation, which was set up in 1965, included not only the low-interest loan business, but also the 'construction and transfer of pollution prevention facility' scheme as well.[16] Under the 'construction and transfer' scheme, complete facilities are transferred at construction cost on the condition of long-term low-interest loans, to relocate factories discharging pollutants away from multiuse areas of residence and industries. This was considered especially cordially by the small and medium enterprises. The targets of the low-interest loans of Environmental Pollution Control Corporation were not only small and medium enterprises but also large enterprises. Most of the loans were financed for large enterprises during the mid-1970s; the new loans to large enterprises in FY1975 reached approximately 100 billion yen. However, in later years, most of the loan customers were small and medium-sized enterprises (Figure 1.3).

The JDB financing for pollution prevention accounted for the biggest share among such loans offered by several government-affiliated financial institutions, and the JDB loans had a heavy impact on the trend in pollution prevention investment by large industries, such as power generation, iron and steel, oil refining and petrochemical, etc.[17] From FY1970, the amount of JDB loans for pollution prevention expanded rapidly. The actual amount of new loans exceeded 200 billion yen in FY1975, setting a record, and it exceeded by 25 per cent the new loans by JDB during FY1974 and FY1976. During this period, pollution prevention loans were one of the main businesses in the JDB lending scheme (Figure 1.4).

The JDB programme provided loans for large-scale plant investment related to pollution prevention, such as the heavy oil desulphurization facility of a petroleum refinery and the LNG (liquefied natural gas) unloading, storage and vaporization facility at a thermal power plant. The heavy oil

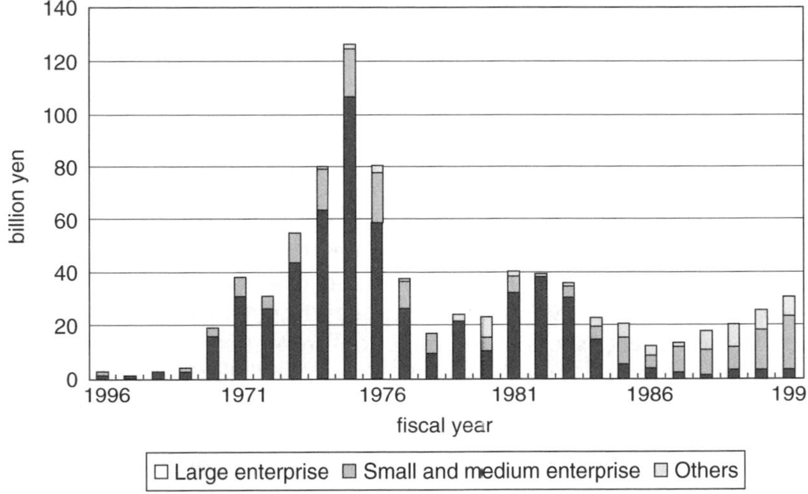

Figure 1.3   Annual loan lending by Pollution Control Service Corporation
Source: Pollution Control service Corporation, *Jigyo Nenpo*, various issues.

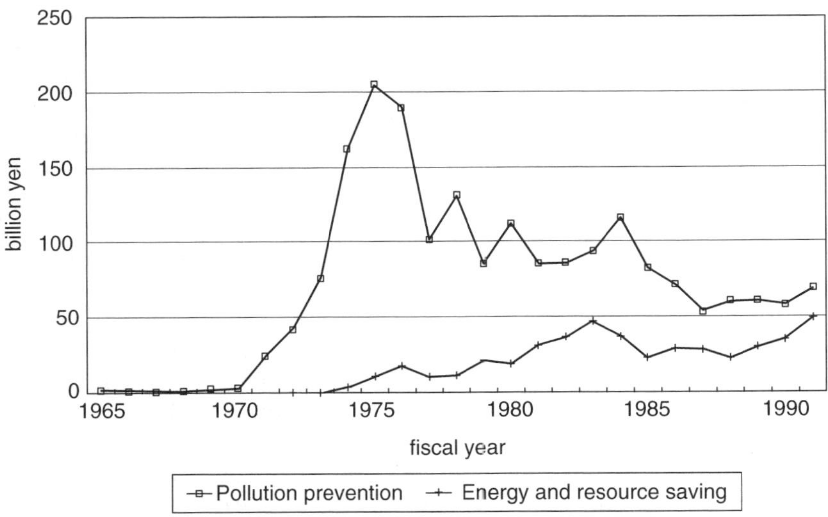

Figure 1.4   Annual loan lending by JDB on pollution prevention and energy saving
Source: Japan Development Bank, *Jigyo Nenpo*, various issues.

desulphurization facility itself is a large-scale plant in an oil refinery. The JDB loans covered investment for the flue gas desulphurization equipment too. Therefore, the loan supported the major part of the 'low sulphurization' plan on financial fund supply side.

Most of the government-affiliated financial institutions and the local governments were involved in policy finance for the small and medium enterprises, and the entire structure of policy finance was quite complicated. Although the financial institutions and their lending facilities partially overlap each other, they each had their own characteristics and were segregated within their demarcations. Most of these financial institutions each had their own loan programme for pollution prevention investment, and each institution's programme reflected the characteristics of that institution.

Japan Finance Corporation for Small Business (JFS) started a lending scheme for the pollution prevention facilities of the small and medium enterprises in FY1965 and took the central role in policy implementation.[18] JFS loans accounted for a large share of the pollution prevention loans for the small and medium enterprises, until the mid-1980s when Pollution Control Service Corporation reduced its loans to large enterprises and began to concentrate on small and medium enterprises (Figure 1.5).

Pollution prevention investment by private enterprise increased rapidly during the latter half of 1960s and the first half of 1970s. At that time, the importance of 'pollution control' as an emergent policy issue led to an expansion of the credit ceiling of the government-affiliated financial institutions.

It is quite difficult to compare the total amount of loans from the government-affiliated financial institutions for pollution prevention investment by private enterprise because the terms and conditions, such as loan-applicable equipment, interest rates, repayment periods and limit amounts, differ markedly among the many loan programmes. When pollution prevention investment by large private enterprises reached its peak in FY1975, the annual total of new loans for pollution prevention loaned by JDB and the Pollution Control Service Corporation amounted to approximately 300 billion yen, which constituted the majority of the loans for large enterprises. According to a survey by MITI, pollution prevention investment by private enterprise(i.e., manufacturing, electric power, gas and mining industries, which were capitalized in excess of 10 billion yen) amounted to a minimum of approximately 1,000 billion yen in FY1975. It can be seen that pollution prevention loans provided by JDB and the Pollution Control Service Corporation covered a significant part of the pollution prevention investment of large private enterprises at that time.

For small and medium enterprises, government-affiliated financial institutions supplied low-interest loans for pollution prevention investment of approximately 40 billion yen in FY1975, when such investment reached its peak. When one includes the loans provided by local governments to small and medium enterprises, which amounted to approximately 40 billion yen, the total amount loaned to small and medium enterprises in FY1975 was approximately 80 billion yen. According to the investment survey by the Japan Small Business Corporation, the total amount of pollution prevention investment by small and medium enterprises in this year was approximately 80 billion yen.

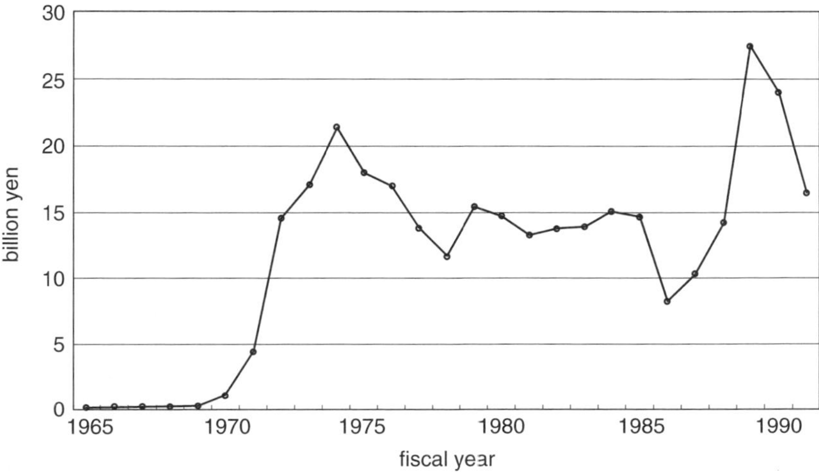

Figure 1.5 Annual loan lending on pollution prevention of small and medium enterprises by JFS
Source: Japan Finance Corporation for Small Business, *Gyomu Hokokusho*, various issues.

This survey indicates that the loans provided by government-affiliated financial institutions for pollution prevention investment by small and medium enterprises covered a very high percentage of private pollution prevention investment, although it may be assumed that pollution prevention investment and the definition of the small and medium enterprises differ at each institution.

Preferential tax measures.   Preferential tax treatment has often been utilized as an indispensable means of implementing industrial policy, together with low-interest loans by government-affiliated financial institutions. The effect of the preferential tax measures is the same as the effect of subsidy payments, so that preferential tax measures are called 'tax expenditures'. However, because budget restrictions are loose compared to the provision of direct subsidies and no report to the Diet is required, budget restrictions are an easy-going policy instrument that policy-makers can use at their own discretion. On the other hand, monitoring of abuse is difficult. Preferential tax measures, once introduced, have often been abused rather than serving their original policy purpose and have sometimes been preserved due to the vested interests of beneficiary groups even after their significance for policy objectives ceased to exist. Unlike with direct subsidy payments, it is difficult to predict in advance how much one preferential measure will cause tax revenue to decrease. The biggest difference between preferential measures and subsidy payments is that subsidy payments are not available to the corporations in deficit.

Most of the preferential treatment measures related to the national tax are based on the Special Taxation Measures Law. Various preferential treatments, such as an 'allowance reserve' that is designated by corporation tax law, also exist in the corporation tax system, in addition to the measures prescribed by Special Taxation Measures Law.

A portion of local taxes on corporations, including the corporate enterprise tax, prefecture residents' tax and local municipal tax, are collected based on the amount of national tax payment (which in case of a corporation is the corporation tax). Therefore, preferential tax measures instituted by the central government for the national tax (the corporation tax) influence the local taxes, too. The decrease in local tax revenue is divided into two parts: the reduction caused by the effects of the central government's measures and the reduction caused by measures directly based on the local tax law.

There are two types of preferential tax measures. One is the 'deduction-type' measure, such as income deductions and tax exemptions, and the other is the 'deferment-type' measure, such as special (initial) depreciation, pollution control reserve funds and allowance revenues.

'Deduction-type' measures decrease the amount of a tax or taxable income directly. One example would be reduced rates for and exemptions from the fixed local property tax. Regarding the 'deferment-type' measures, such as special depreciation, reserve funds and allowance reserves, we briefly explain in the following section.

Special depreciation measures accelerate depreciation of property in corporate accounting under some methods and enable the depreciation to be entered under losses. This is done to reduce the corporation's book profit, which is the basis for the computation of the amount of the corporation tax to be paid, thereby reducing the corporation tax. Special depreciation measures can postpone corporate tax payment to the future by taking the future amount of depreciation in advance. Therefore, a special depreciation measure is not a direct tax deduction measure such as a tax credit or tax deduction from income, but only has the effect of postponing taxation to future years. So long as the corporation tax payment is usually delayed in comparison with the repayment, a special depreciation has an effect similar to getting an interest-free loan.

Reserve fund and allowance reserve measures permit an enterprise to keep a tax-free reserve fund within a fixed period and up to the ceiling, when there are some profits, to prepare for future payments and hedge against the risk of income fluctuation in future. It becomes taxable as part of the corporation tax when it is withdrawn from the reserve fund. The reserve fund measure has the same effect as an interest-free loan for capital accumulation by enterprises, because these measures postpone the corporation tax payment, similar to the effect of special depreciation measures.

Together with the fiscal investment and loan programme, the preferential tax treatments, such as special depreciation, reserve funds and allowance reserves, occupied a very important position in the industrial policy of post-war Japan. Private enterprises emphasized preferential tax measures as an effective means of capital accumulation, and preferential tax treatment was an important means of stimulating plant investment in the direction favoured by government policies.[19] Because the actual effective interest rate was high enough throughout the 'high growth period' in Japan, it seems likely that the effect of the interest-free loans provided by preferential tax measures for plant investment was strong in that period.[20]

As for the preferential tax treatment related to the pollution prevention investment, there are provisions in the Basic Law for Environmental Pollution Control, Air Pollution Control Law and Clean Water Law. As a measure for pollution prevention in the national tax, the corporation tax was decreased through special depreciation measures and a pollution control reserve fund based on the Special Taxation Measures Law.[21]

Looking at the decreases in tax revenue amount, the exemption from or reduction of the fixed property tax was the most important preferential measure at the local tax level, except for the decrease in local taxes along with the decrease of corporation tax implemented through the preferential tax measure of the central government.[22]

The decrease in tax revenue (or for the enterprises, the amount of tax payment reduction) through the special taxation measures related to prevention of pollution is shown in Figure 1.6. The overall decrease in tax revenue was estimated at approximately 100 billion yen at its peak in FY1975. National tax (in this case, corporate tax) revenue reduction related to pollution prevention alone reached approximately 60 billion yen, or approximately 20 per cent of the whole corporate tax decrease of FY1975. These data were estimated with a budget base by the National Tax Bureau of the Ministry of Finance, and reported to the Government Tax Commission (the Prime Minister's consultative council).

A special depreciation measure by the central government can give an economic incentive to private enterprises to investment in pollution prevention equipment. Pollution prevention investment in most cases does not directly cause profit expansion for the enterprise concerned. Therefore, when the rapid execution of pollution prevention is required, it is not easy for the private enterprises to recover their investment costs incurred for pollution prevention equipment in the short run.

Special depreciation measures applied to the pollution prevention facilities can trigger pollution prevention investment because the burden on the enterprise to carry out pollution prevention investment is clearly reduced so long as the measure is equivalent to an interest-free loan.[23] When special depreciation measures for pollution prevention investment were introduced in FY1967, the ratio of special depreciation deduction for the first year after pollution prevention investment was set at 33 per cent and was raised to 50 per cent in FY1971. It was reduced to 33 per cent in FY 1977 and then decreased continuously to reach 27 per cent in FY1980, 25 per cent in FY1982 and 22 per cent in FY1985. From FY1987, it came to be accepted only when pollution prevention equipment was set up along with new construction of a plant.

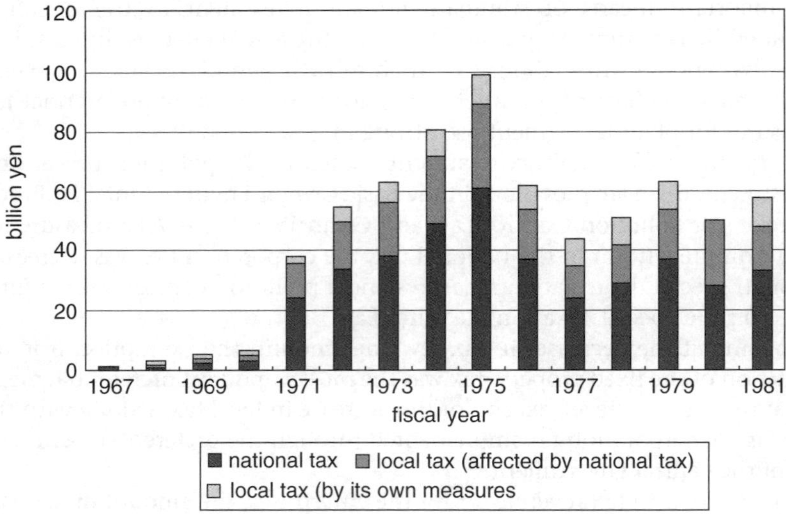

Figure 1.6   Decrease in tax revenue by special tax treatments on pollution prevention
Source: Government Tax Commission, *Zeisei Chosakai Kankei Shiryoshu*, various issues.

The pollution prevention reserve fund was introduced in FY1972, to avoid placing too large a burden on private enterprises by having them pay for pollution prevention cost when their business performance was declining. The limit on the additional annual reserve amount was established as 0.3 per cent or 0.6 per cent of the sales revenue, depending on the type of industry. Even if more pollution prevention investment was made, there was no benefit earned by the enterprise under this system. The pollution control reserve did not provide any incentive to private enterprises for pollution prevention investment because the purpose of the reserve was to provide for the operation cost of pollution prevention facilities.

According to Tadashi Murai, the pollution prevention reserve fund was a unique system in Japan. The pollution prevention reserve fund system did not provide an incentive for enterprises to exert more effort to prevent environmental pollution. Furthermore, when payments were made from the pollution prevention reserve, there was no investigation by the government to confirm that the payments were actually utilized to cover the pollution prevention costs of the enterprise. The pollution prevention reserve was criticized, in the National Assembly right after its introduction, for taking on the characteristics of a means of profit retention in the name of pollution prevention.[24] The National Tax Agency was forced to admit the necessity of a re-examination of the reserve in FY1974, only two years after its introduction. At that time, the sales percentage that could be added to the reserve fund was lowered, from 0.6 per cent to 0.3 per cent and from 0.3 per cent to

0.15 per cent, depending on the type of industry. Finally, in FY1978, it was decided to phase out the reserve after a transitional period of three years.

At the local tax level, the majority of the tax revenue reduction related to pollution prevention is due to the exemption from and reduction of the fixed property taxes, together with the preferential measures. The history of change in the rate of reduction and the coverage of specific equipment almost corresponds to the historical transitions in the special depreciation measure of the national tax related to pollution prevention. The exemption from and reduction of the fixed property tax causes a tax decrease, not only on the new pollution prevention investment but also on the fixed assets acquired by past pollution prevention investment. Therefore, its effect is to induce new pollution prevention investment, but only indirectly by reducing the long-term cost of pollution prevention of the enterprise, rather than as a direct incentive for new pollution prevention investment.

Apart from preferential tax measures relating to pollution prevention, which we have already discussed, there is another type of tax treatment to promote pollution prevention efforts by private enterprises indirectly. A pollution tax or surcharge on pollution emission can induce pollution prevention investment by private enterprises, similar to preferential treatments, by increasing the cost to the enterprises of emitting pollution into the environment. According to microeconomic theory, surcharges on pollution emission and subsidies for pollution prevention should, at least in the short run, have the same effect as incentives to private enterprises for pollution prevention. However, subsidies are not a desirable method from the viewpoint of 'the polluter pays principle' (PPP). A surcharge is more desirable than a subsidy because a subsidy encourages the polluting industry in the long run. Nevertheless, it was only after global environmental problems become a policy issue in the latter half of the 1980s that the introduction of taxes came to be examined realistically in Japan. Until then, only various preferential tax treatments had been instituted in Japan as tax system measures for the prevention of pollution.

Preferential tax treatment for pollution prevention is an effective means of providing incentive for pollution prevention activities to private enterprises, and preferential tax treatment can be considered one of the factors that led to an enormous amount of pollution prevention investment in Japan within short period during the mid-1970s. However, many of those preferential tax measures remained in effect even after intensive pollution prevention investment was accomplished and the annual amount of investment had decreased. A review of the preferential tax treatment system related to the plant investment as a whole started from the end of the 1970s, and many parts of the system related to pollution prevention were abolished during the process. The amount of the corporation tax revenue exemption due to preferential tax treatment in the budget base decreased rapidly in the 1970s. However, the abolition of preferential treatment related to pollution prevention was delayed compared

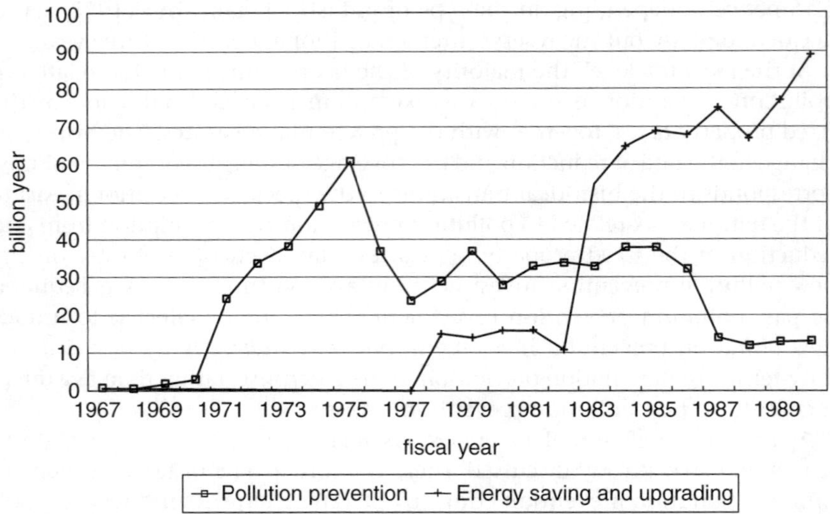

Figure 1.7   Decrease in national tax by special tax treatments on pollution prevention and energy saving
Source: Government Tax Commission, *Zeisei Chosakai Kankei Shiryoshu*, various issues.

to the abolition of other preferential treatments. This was often pointed out as evidence of an unfair taxation system. Following that, an overall check of the preferential treatments finally was undertaken in the latter half of the 1980s.

While preferential tax treatments related to pollution prevention were abolished gradually, the preferential tax measures related to energy saving and the promotion of alternative energy use were expanded rapidly after FY1981. From among the corporate tax revenue decreases caused by preferential tax measures, we compared annual trends in the amounts related to pollution prevention and the amounts related to energy saving (including upgrading of energy use) in Figure 1.7. A sudden expansion in the amount of corporate tax revenue decrease affected by energy use reflects the fact that special taxation measures related to the energy use included very powerful measures for tax credit from FY1984. Tax credit, which is a 'deduction-type' measure similar to a deduction from taxable income, is a much more powerful preferential treatment than the 'deferment-type' measures, such as special depreciation, reserve funds and allowance reserves. All of the significant 'deduction-type' measures for corporate tax had already been abolished in the 1960s.

As far as pollution prevention through special taxation treatments is concerned, the 'deduction-type' measure at the national tax level, such as tax credit and deduction from taxable income, has never been accepted by the National Tax Administration Agency. Since the mid-1980s, special taxation treatment related to energy use has been notable for the intensity and strength of its measures. It is said that historical trends in new establishment

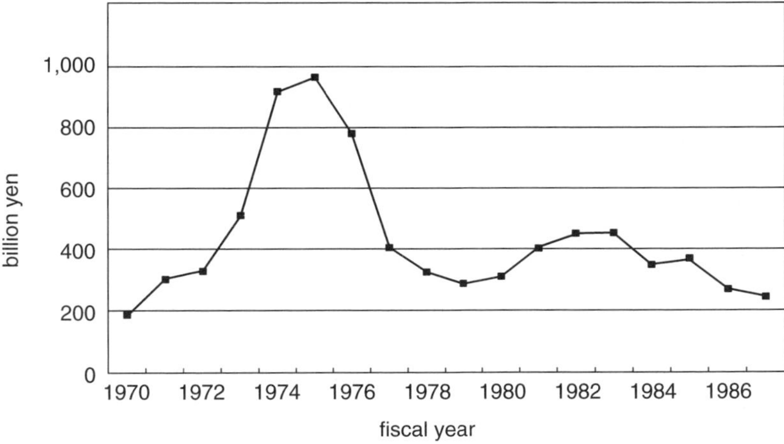

Figure 1.8 Pollution prevention investment by major private manufacturing enterprises
Source: Ministry of International Trade and Industry, *Shuyo Sangyo no Setsubi Toshi Keikaku*, various issues.

and abolishment of various preferential tax measures reflect the shifting emphasis in industrial policy, which is shifting from pollution prevention to energy conservation and new energy development. Moreover, manufacturing industry's energy-saving investment displays a close relationship with pollution prevention activities (especially air pollution). Some types of investment, such as that for improving the efficiency of boilers, could be defined as either type of investment, and so it may be possible to change the classification of investments from pollution prevention to energy saving. Although most of the special taxation measures related to pollution prevention have been abolished, some were incorporated into more powerful measures related to energy saving, thereby prolonging their lives substantially.

Pollution prevention investment by major private manufacturing enterprises, including electric power companies, is shown on Figure 1.8. At its peak in FY1975, total amount of private pollution control investment in Japan was about a trillion Japanese yen. It was about 17.7 per cent of total capital investment of those enterprises in FY1975.

3.   Energy policy and industrial pollution control: political economy of 'low sulphurization'

An examination of energy policy is important when considering the relationship between industrial policy and industrial pollution regulation, since energy policy is related to both industrial policy and industrial pollution control.

Energy policy in particular has a deep relation with industrial air pollution. For example, skyrocketing energy prices in the 1970s worked as a powerful incentive for private enterprises to reduce energy use, which also reduced emission of air pollution from industries. It can be said that these energy conservation efforts brought about a reduction in the discharge of air pollutants, such as SOx which occurs in direct proportional relation to the sulphur content in the energy consumed. Below, in order to show the importance of the relationship between energy policy and measures to reduce industrial pollution, we will focus on the example of 'measures for low sulphurization' and energy conservation.

Among the 'industrial pollution regulations for achieving industrial policy', it can be said that SOx regulations achieved some degree of successful outcome by means of a 'combination of direct regulation and preferential treatment'. We will consider the implications of the measures in air pollution prevention policy to reduce SOx by focusing on the relationship with energy policy.

The central part of the SOx regulation policy from the middle of the 1960s was the reduction of discharge from the fixed sources, such as factories and power generation plants. The concrete contents of a series of regulations called the 'low sulphurization measures' were (1) low sulphurization of imported crude oil; (2) installation of heavy oil desulphurization facilities in oil refinery plants; (3) installation of stack gas desulphurization facilities; and (4) heightening of smokestacks, etc.[25] The source of the SOx discharged into the atmosphere was the sulphur content in fuel, and since low sulphurization of fuel was the easiest countermeasure technologically and financially, it was first pursued as much as possible as a measure against SOx pollution. However, fuel gas desulphurization was not yet established technologically or commercially in the mid-1960s, so it was not to be relied upon at that time.

The legal basis of the measures against SOx was the Clean Air Act, which was enacted in 1962 and amended and expanded in 1968. However, administrative guidance without legal basis was serially issued to the industrial community (on the supply side and demand side of energy) behind the scenes. The Advisory Committee for Energy, a consultative body of the Minister of International Trade and Industry, prepared the committee report 'An Optimal Comprehensive Energy Policy' in February 1967 and took up the 'low sulphurization plan' as a subject. Furthermore, the Committee established a 'Subcommittee on Measures for Low Sulphurization' in May 1969, and it summarized the 'low sulphurization plan' in December of the same year. To carry out 'low sulphurization' effectively and extensively, the plan set forth long-term guidelines to show industries that used fuel how to set up targets for fuel use and sulphur content for each region, as well as guidelines for oil refineries that supply heavy fuel oil, which indicated targets for average sulphur content and production amount for fuel in each oil classification. Based on this plan, administrative guidance to both the supply side and the demand side was conducted intensively.

Although importation of crude oil with lower sulphur content was an effective measure, the tightness of the market after the first oil shock made it difficult to continue such importation. Thereafter, installation of the heavy oil desulphurization facilities by oil refinery companies was conducted, followed by the installation of the stack gas desulphurization facilities by industries on the fuel demand side.

On the other hand, in order to carry out this plan, the government promoted the technical development of heavy oil desulphurization facilities and stack gas desulphurization facilities, as well as installation of the equipment for these facilities. The financial burden placed on enterprises that carried out those measures was alleviated by low-interest loans from government financial agencies, including JDB, and by accelerated depreciation through special taxation measures, exemption from and mitigation of fixed property tax, and mitigation of import duties on low-sulphur crude, etc.

In the process of executing the 'low sulphurization measures', MITI issued administrative guidance, in connection with the adjustment of interests among related industry circles, and asked for cooperation among them. The related industries were the electric power and the iron and steel industries, etc., which are large users of heavy oil, and the oil refinery industry, which is the supplier. In addition to fuel, the oil refining process produces naphtha in a fixed proportion, which serves as a raw material mainly in the petrochemical industry. Therefore, the trend in heavy oil demand was a matter of serious concern for the petrochemical industry as well because it requires naphtha as a raw material. These industries were under the powerful influence of MITI at that time. Through the 'Petroleum Act' enacted in 1962, oil refineries in particular were the targets of powerful intervention by MITI regarding their quantity of production, prices, and plant and equipment investment, etc.[26]

Prior to the 'low sulphurization plan', the electric power industry had started combustion of crude oil, which had already been substituted for heavy oil as fuel for thermal power plants starting in 1962. The sulphur content of heavy oil is higher than that of crude oil because most sulphur remains in the heavy oil in the process of refining. However, the electric power industry started crude-oil combustion in thermal power plants not because the sulphur content of crude oil is lower than that of heavy oil, but because the price of the heavy oil refined in Japan was relatively high. The prices of petroleum products are actually controlled, with the price of naphtha fixed low for the promotion of the relatively new petrochemical industry and the price of heavy oil set comparatively high.

Moreover, it was more advantageous for the electric power industry to limit measures against SOx to fuel measures and to put off introduction of the stack gas desulphurization facilities in thermal power plants as long as possible. Fuel gas desulphurization was not yet technologically viable in the late 1960s. For the electric power industry, the deadlock in measures to reduce SOx through fuel low sulphurization meant that the cost of the emission gas desulphurization, which had not been fully established technically, had to be paid.

Since the petroleum policy of MITI was premised on adherence to the 'principle of refining near the consumption region', and the pattern and organization of the oil refinery industry at that time, heavy oil was produced from crude oil at a fixed rate technically. Therefore, in order to maintain the production quantity of lightweight oil, including naphtha, which was in tight supply in those days, the oil refinery companies had to secure the domestic market for heavy oil. Moreover, if crude oil was not refined and combusted in electric power plants as a fuel, it would bring no profit to the oil refinery industry. Therefore, the oil refinery industry was opposed to expansion of crude-oil combustion by the electric power industry. Combustion of low-sulphur crude oil would suppress the supply of low-sulphur crude for refinery industry, and the oil refinery industry insisted that this was not an efficient way to pursue low-sulphurization measures in Japan as a whole. A report in February 1967 issued by the Advisory Committee for Energy insisted that crude-oil combustion by the electric power industry should be 'within the limits of the type C heavy oil import schedules amount', in response to the objections of the oil refinery industry.[27] However, later this limit was clipped away gradually in increments.

Although continued implementation of pollution regulations would generally cause cost increases for the electric power industry, the social imperative of 'low sulphurization' provided the electrical power industry opportunities to escape the restrictions on fuel use that the government and the domestic fuel industries imposed, to realize diversification of fuels for thermal power generation, to reduce use of the high-priced domestic heavy oil, which was required by the government, and to reduce costs simultaneously.

Moreover, the electric power industry was assigned to receive the 'Kafji crude oil', which the Arabian Oil Company, a national policy concern, developed in those days. When crude-oil combustion began at electric power plants, Kafji crude oil was burned. However, the high sulphur content of Kafji crude oil became a problem, and the electric power industry requested a reduction in the amount of Kafji crude oil that it received.

The interest of the petrochemical industry in the 'low sulphurization measures' was almost the same as the interest of the oil refinery industry. Crude-oil combustion by power plants without separating the parts of lightweight crude oil, such as naphtha, could undermine the stable supply of naphtha for industrial use, and this was a serious obstacle to the development of petrochemical industry.

With such interests among related industries in the background, the 'low sulphurization measures' were promoted by the government. An important turning point was brought about by the 'pollution control agreements', which are a means of pollution regulation used by local governments. The

Tokyo metropolitan government and the Tokyo Electric Power Co. (TEPCO) concluded a 'pollution prevention agreement' in September 1968, on the occasion of the construction of a thermal power plant on the Oi wharf reclaimed land. In the agreement, TEPCO promised to use only 'Minas crude oil' with super-low sulphur content as fuel from FY1973 in the Oi Thermal Power Plant. Exclusive usage of Minas crude oil was proposed by TEPCO in the process of negotiation. The sulphur content of Minas crude oil was only 0.1 per cent, which was lower than the 1.7 per cent technical limit on sulphur content at the heavy oil desulphurization facilities (which used an indirect desulphurization method) in those days. Furthermore, TEPCO made a firm promise to continue using super-low-sulphur crude oil in the future at the Oi Thermal Power Plant.[28] The conclusion of this pollution prevention agreement was a big shock for the oil import, refinery and petrochemical industries. Both the oil refinery and petrochemical industries were opposed to combustion of super-low-sulphur crude oil, for which the absolute quantity was restricted, at electric power plants without refining. In the process of negotiation between TEPCO and the Tokyo metropolitan government, there was no trace of involvement by MITI.[29]

Although, in those days, the local governments did not have the legal authority to implement pollution regulations that were severer than those of the central government, to cope with the conditions at that time, the local governments needed a regulatory means to deal with the actual conditions of their areas, since the standards of the central government's pollution regulations were too loose. 'Pollution prevention agreements' were invented under such conditions, and they were extensively utilized as a means of pollution control by most of the local governments in Japan.[30] The Yokohama Municipal Government concluded one of the very early examples of such an agreement with the Electric Power Development Co. Ltd, in 1964. It is known that the strong pressure from the residents in each area served as an impetus for local governments, such as the Yokohama municipal government, to press the enterprises in the area to conclude pollution prevention agreements. For the enterprises, the pollution prevention agreements were also a promise to the local community through the local government regarding prevention of pollution. Moreover, it was more desirable for the enterprises to prevent disputes with residents by concluding agreements with local government, thereby clearly showing their attitude towards pollution prevention to local residents, rather than risking a dispute later by leaving the possibility of pollution generation ambiguous. Without a definite promise on pollution prevention, neither local governments nor residents in mixed-use residential and industrial areas could accept new factories in their areas.[31] At most of TEPCO's thermal power plants, effective SOx emission control was realized through strict pollution prevention agreements.

The electric power industry was able to advantageously advance negotiations on combustion of crude oil with the oil refinery industry and the petrochemical industry, and with MITI, using as leverage the pressure from local governments that were responding to their local residents' demands for pollution prevention. After 1968 when TEPCO concluded its first pollution control agreement with the Tokyo metropolitan government, the nine major electric power companies in Japan successively signed pollution prevention agreements with local governments where their plants were located. Most agreements included specific regulations on low sulphurization of fuel.[32]

Furthermore, the electric power industry, together with the iron and steel industry, began using more naphtha as a low-sulphur fuel in 1971. At that time, a cloud hung over demand expansion of petrochemicals. Tightness in the supply of naphtha was easing, and a surplus of naphtha was seen overseas. Therefore, these industries could expect to procure a certain amount of naphtha at relatively low prices. Needless to say, naphtha combustion by electric power plants and the iron and steel industry stimulated the petrochemical industry more than the oil refinery industry.[33] Although the oil refinery and petrochemical industries tried to oppose such usage, naphtha combustion by electric power plants and the iron and steel industry was accepted after all.

In order not to give the electric power industry any more excuses to expand crude oil combustion under such circumstances, the oil refinery industry was forced to expedite desulphurization of heavy oil. Introduction of heavy oil desulphurization facilities at oil refineries, as well as technological developments and improvements, were advanced very quickly.

Heavy oil desulphurization facilities were already afflicted by a low operation ratio caused by over-capacity in the first half of the 1970s. As operation ratio falls, cost per unit of production increases markedly. The causes of over-investment were that (1) heavy oil desulphurization alone was inadequate because SOx emission regulations were tightened; (2) the electric power industry, which is a large user, coped with the tighter regulations by combusting crude oil and naphtha at power plants; and (3) heavy oil desulphurization was emphasized more than necessary in the 'low sulphurization plan'.[34] If the oil refinery industry was going to prevent expansion of crude oil and naphtha combustion, it had to expedite desulphurization of heavy oil. This is considered to have been a cause of the over-investment in heavy oil desulphurization facilities. The capacity of a heavy oil desulphurization facility is shown in Figure 1.9. It turns out that there was hardly any increase in capacity from the second half of the 1970s.

Anti-pollution investment at that time by oil refineries is shown in Figure 1.10. It turns out that the majority of investment was spent on heavy oil desulphurization facilities (i.e., equipment to prevent pollution by consumption of the product). Therefore, the investment in the equipment to prevent pollution by the oil refinery plant itself was not as large as it appears on the whole. Although the operation ratios of the heavy oil desulphurization facilities were already very low in 1975, large-scale investment was continued until around FY1976.

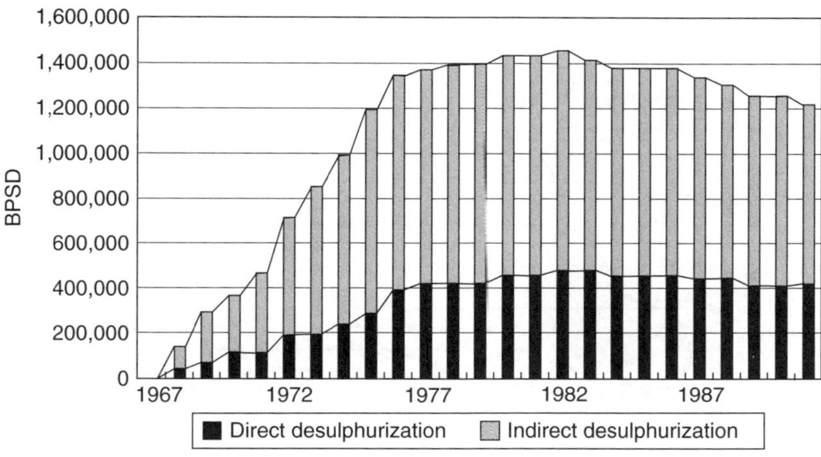

Figure 1.9   Capacity of heavy oil desulphurization facilities
Note: Capacity at the end of each year.
Source: Petroleum Association of Japan, *Naigai Sekiyu Shiryc: FY 1991 edition*, 1992.

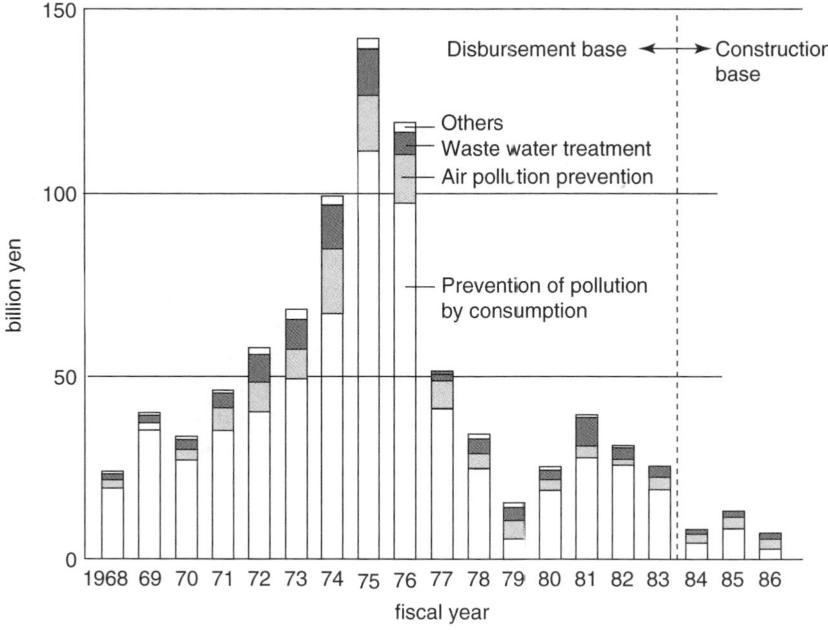

Figure 1.10   Pollution prevention investment by oil refineries
Source: Petroleum Association of Japan, *Sekiyu Renmei Shiryo*.

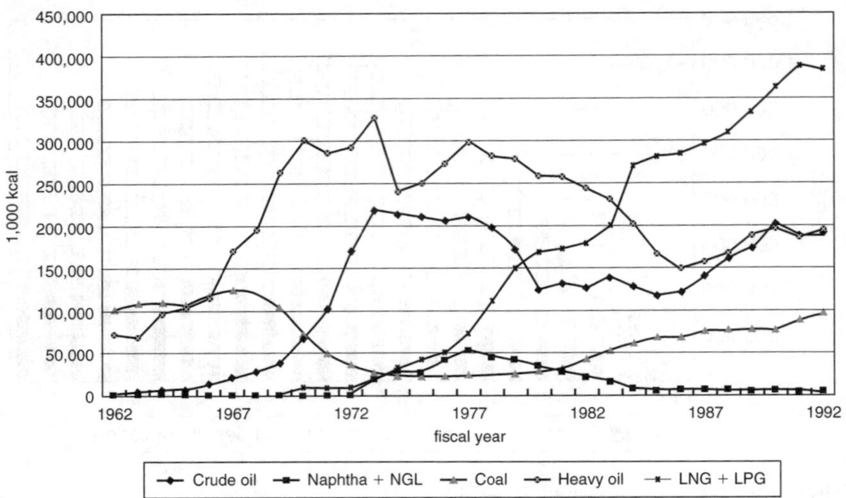

Figure 1.11   Fuel use of thermal power plants of nine major electric companies (calorific power base)
Sources: Federation of Electric Power Companies of Japan, Denkijigyo Binran; Agency of Natural Resources and Energy, *Sogo Enerugi Tokei*, various issues.

Shown in Figure 1.11 is the amount of each fuel used at thermal power plants of the nine major electric power companies from 1962, when crude-oil combustion started at power plants. Crude-oil combustion rapidly expanded with the 'low sulphurization plan', and the amount consumed did not fall off as with heavy oil, even after the first oil crisis. Although coal use decreased rapidly in the late 1960s, it recovered considerably after the second oil crisis in 1979. LNG continued rapid growth, starting in the 1970s, and became the mainstream fuel of thermal power plants from the middle of the 1980s. LNG requires transportation and storage facilities for its own exclusive use and, although it is high cost, it has no sulphur content at all. With thermal power plants being newly located around urban areas, the only way to meet the strict requirements of SOx emission control is to construct power plants that use LNG exclusively.

When the interests of the oil refinery, petrochemical and electric power industries were opposed to each other in the 'low sulphurization measures', all that MITI did was to coordinate the opinions of the *Genkyoku* (the sections of the bureaucracy within the government that had the primary responsibility for development and supervision policies for specific industries), which were located within MITI. It seems that MITI was unable to set forth an energy policy that went beyond the adjustment of interests among the industries to make the pollution control measures more compatible. Moreover, as a matter of petroleum policy, administrative guidance seldom took into consideration the effects of substitutions for fuel and materials.

This caused serious conflicts of interest, as seen in the relation between crude oil and naphtha combustion and low-sulphur heavy oil.[35] However, air pollution prevention measures taken by these industries progressed eventually as requests were met concerning crude oil and naphtha combustion by the electric power and the iron and steel industries. In fact, it can be said that the social need for prevention of air pollution brought about the de facto fall of the authority of MITI over fuel use by the electric power industry.

It can be said that the 'low sulphurization measures' scored a success as an industrial pollution regulation due to administrative guidance on fuel (except in the cases of large fuel users, such as the electric power, iron and steel, and petrochemical industries mentioned above). The concentration of SOx fell rapidly. This contrasts with the nitrogen dioxide concentration, which did not improve enough. The combination of 'direct regulation' and 'preferential treatment' using administrative guidance and experience with policies for promoting industrial growth was the easiest method for the policy authorities, because the same means they used for promoting policies for industrial growth prior to industrial pollution regulation could be applied in this case as well.

4. 'Developmentalism' and political economy of industrial pollution

'Developmentalism' is based on the principle of a market economy that maintains market competition, a private property right system and trust in the mechanism of efficient resource allocation through fair market exchanges. As we have already seen in the previous section, 'developmentalism' will hopefully be able to effect rapid industrialization by enjoying the 'advantage of backwardness'. The faster industrialization is realized, the higher the level of industrial pollution corresponding to the high growth, and that could cause the serious damage to the environment and human health if proper countermeasures are not carried out simultaneously. However, 'developmentalism' could create countermeasures very rapidly if industrial pollution problems are recognized and pollution control regulation is incorporated in the system of development policy.

Chalmers Johnson referred to the pollution control policy of Japan (a 'plan-oriented system' in his terminology), comparing it with the policy of the US (a 'market-rational system'), in his well-known book on Japanese industrial policy, as follows:

> [The] most important evaluative standard in market rationality is 'efficiency.' But in plan rationality this takes lower precedence than 'effectiveness.' Both Americans and Japanese tend to get the meanings of efficiency and effectiveness mixed up. Americans often and understandably criticize their official bureaucracy for its inefficiency, failing to note that efficiency

is not a good evaluative standard for bureaucracy. Effectiveness is the proper standard of evaluation of goal-oriented strategic activities.[36]

Both types of systems are concerned with 'externalities', or what Milton Friedman has called 'neighborhood effects' – an example would be the unpriced social costs of production such as pollution. In this instance, however, the plan-oriented system has much greater difficulty than the market-rational system in identifying and shifting its sights to respond to effects external to the national goal. The position of the plan-rational system is like that of a military organization: a general is judged by whether he wins or loses. It would be good if he would also employ an economy of violence (be efficient), but that is not as important as results. Accordingly, Japan persisted with high-speed industrial growth long after the evidence of very serious environmental damage had become common knowledge. On the other hand, when the plan-rational system finally shifts its goals to give priority to a problem such as industrial pollution, it will commonly be more effective than the market-rational system, as can be seen in the comparison between the Japanese and American handling of pollution in the 1970s.[37]

As rapid industrialization boosted the national economy and income distribution equalization expanded the level of consumption, the standard of living was finally improved significantly. Then, most people eventually turned their attention to preservation and improvement of their ordinary lives and living environment and later on raised concerns over the preservation of the natural environment. However, a time lag can be expected between emergence of the pollution problem and social recognition of the seriousness of its damage and discomfort, so that substantial time is required to prepare the establishment of an institutional framework to prevent pollution and to promote efforts to conserve the natural environment. The delay of countermeasures caused by such a time lag placed an excessive load on environmental resources, which originally were renewable, and may have caused irreversible destruction to them.

Even after industrial pollution has been recognized as a social problem, no movement is made towards a solution if the matter is entrusted only to the market mechanism. Industrial pollution problems are an example of the inefficiency of 'market failure', created by negative 'externality'. The negative externality could be removed by 'voluntary negotiation' between the parties concerned (polluters and victims in the case of pollution) in the market economy based on the private property rights system. The problem of negative externality must be left unsolved when the 'transaction cost' to organize 'voluntary negotiation' is larger than the possible profit from an agreement achieved through such negotiation. As for the industrial pollution problem, we could imagine many cases where the transaction cost to achieve a voluntary solution through negotiation is prohibitively high so as to preclude possible negotiation between parties concerned. For example, even if damage due to the pollution is already recognized by society, the cause-and-effect relation must also be recognized to realize compensation. The cost of organizing victims could be enormously high if pollution spreads across a wide area and the number of victims is large.[38] In the case that solution through voluntary negotiation is difficult, there is a possibility that government intervention could improve the allocation of resources and enhance social welfare. In the case of environmental resources, the optimum allocation of resources will not necessarily be achieved if left to the market mechanism. Policy intervention is necessary to improve allocation of resources and to enhance social welfare by making up for the deficiencies of the market mechanism.

Japan could achieve rapid economic growth by using social and administrative structure to pursue economic growth and by using the policy system based on 'developmentalism' that existed in Japan. As an inevitable corollary of such economic success, environmental problems at first appeared and then were recognized by the people and the government as an industrial pollution problem. The government and private enterprises were pressed to adopt countermeasures by the discontent and protests of the people over the industrial pollution problem. In adopting countermeasures, a style of regulation that includes direct negotiations on the details of individual pollution prevention measures was selected in view of the relationships between the government and the enterprises. These relationships had been formed through experience in implementing industrial policy that already existed at that time.

The decision-making process for the industrial policy was organized industry by industry, where the bureaucrats of the *Genkyoku* and the representative of each industry played dominant roles. Interests of each industry and producers' interests were preferentially taken into consideration, disregarding the interests of other parties, such as consumers and local residents, who were affected by the development of industry and industrial policy itself to some extent.

Industrial pollution regulation was incorporated as a part of series of negotiations between government authorities and the industries or individual firms. Although most of the industrial pollution regulations had a legal basis, in many cases the authorities relied on 'administrative guidance', which does not necessarily have a legal basis, issued to the industries or individual firms concerning specific measures. Preferential treatment such as low-interest loans and preferential taxation measures, etc., were combined with those regulations through administrative guidance. The combination of 'direct regulation and preferential treatment' was a typical measure in industrial policy. In the situation of Japan of that time where industrial pollution had already intensified, in order to promote measures against industrial pollution and reduce it as fast as possible, it may be said that the policy was effective and necessary, at least in the short run. However, when planning ways to exploit the environmental resources, including land resources, of an economy as a whole over the long run, the limitations of responses through industrial policy or administrative guidance are clear.

The industrial policy that used administrative guidance was an effective policy instrument in the policy system based on 'developmentalism' for giving priority to industrial development above all else. However, the most serious problem of administrative guidance is that the process of negotiation between government and industry or enterprise is done behind closed doors and no record or document is left behind; in addition, the locus of responsibility for the results is ambiguous. Information disclosure is also an important issue in administrative guidance that deals with industrial pollution regulation.

According to an OECD review of environmental policy in Japan published in 1994:

> [the] public should be given access to environmental information and data, including information on voluntary agreements between any levels of government and industry, and information on emission registers; exceptions to this general principle should be limited to defined circumstances. Various forms of environmental reporting should be developed by private enterprises. Public awareness of and participation in decision making concerning the environment should be enhanced.[39]

Based on the Japanese experience, the influence that 'developmentalism' exerts on the appearance of industrial pollution and the response of society to the problem could be summarized as follows.

1. Rapid industrialization guided by industrial policy increased the burden placed on the environment, advanced environmental destruction and pollution quickly, and caused a delay of the countermeasures.
2. With delayed recognition of social problems such as environmental pollution, the industrialization promoted by un-decentralized decision-making was claptrap and brought about only ad hoc and after-the-fact measures.
3. The direct command and control regulations utilizing instruments of the industrial policy were effective at least in the short run, as measures against pollution, although the regulations were limited to after-the-fact measures.
4. There was a side effect in which the certain degree of success of the 'industrial pollution measures' prevented formation of a decentralized decision-making system required for the fundamental solution of environmental problems and amenity improvement, which should include the institutionalization of environmental impact assessments and citizen

participation in the long-term use plan for environmental resources, including national land.

Although the concept of 'developmentalism' is based on the experience of industrial policy and rapid economic development in Japan, even in developing countries to which the formulization of 'developmentalism' does not necessarily apply, it is difficult to find an example where industrialization by the government through a certain type of industrial policy was not tried once. Even if an industrial policy does not succeed in industrialization, environmental destruction will certainly be left behind, in the form of development attempted in vain.

The position of the Japanese experience as an example of industrialization provides an important viewpoint for late-comer countries that allows them to consider generation of and countermeasures for industrial pollution. In industrialization based on 'developmentalism', industrial policy was the most important means. In the Japanese experience, although industrial policy was accompanied by strong governmental intervention, the government never attempted to run a controlled economy under powerful regulation. Instead, the government did fundamentally limit its role to corrective coordination of the market economy. The problem is that, in measures against industrial pollution, it became difficult to form policy or institutions that fully reflected citizens' preferences and interests because the conventional measures used for industrial policy, such as adjustment and negotiation between the government and individual firms or industry, were to be preserved.

We have to be careful if we attempt to apply the 'Japanese experience' of industrial pollution control and environmental policies to developing countries in the present world. The social, institutional and political background, as well as the international environment, policy instruments and technological knowledge are quite different in developing countries at present compared to Japan in its high-growth era. However, we believe the 'Japanese experience' of industrial pollution control could be, at least, a reference or starting point for mutual discussion on how a policy system for environmental resource conservation should be developed during the process of economic growth when strong government intervention is inevitably required.

Notes

1. See Murakami (1996) on 'developmentalism' and economic theory of decreasing marginal cost of industry. Suehiro (1998) also explained concept of 'developmentalism' pursued by developing countries.
2. Hara (1992), p. 159.
3. Komiya *et al.* (1988), Introduction.
4. Baba (1988).
5. Miwa (1988).

6. As for excessive competition, see Okimoto (1989), pp. 38–48, and also Murakami (1996).
7. Chapter 4, 'Tekko: Kakaku to karuteru', in Imai (1976), pp. 127–81.
8. Tsuruta (1988).
9. See Chapter 4, Imai (1976). However, Miwa (1988) doubts the existence of such influence of policy intervention on enterprise behaviour.
10. On typology of industrial policy based on microeconomic theory, see Suzumura and Okuno (1993).
11. OECD (1977), p. 33.
12. OECD (1977), p. 35.
13. OECD (1977), p. 36.
14. Nakaoka (1974), p. 212.
15. Murai (1975), pp. 1–5.
16. Pollution Control Service Corporation (1991) explains details of its operation.
17. On its loan programme for pollution prevention investment and its actual lending performance, see Japan Development Bank (1976), pp. 327–43.
18. Japan Finance Corporation for Small and Medium Business (1984), pp. 200–1, 302–4.
19. Ogura and Yoshino (1988) and Ikemoto *et al.* (1984); on preferential tax treatment concerning pollution prevention, see Niizawa (1997) and Lee (2004).
20. Tsuruta (1982), pp. 76–8.
21. Murai (1975), pp. 1–3.
22. Wada (1992), pp. 84–6.
23. Murai (1975), pp. 87–90.
24. Murai (1975), pp. 7–9, 11, 100–103.
25. For the report on 'low sulphurization measures' summarized by MITI, see Furuichi (1979) and Yamaguchi (2002). See also Teranishi (1993) and Weidner (1995) on the SOx pollution control policy in Japan.
26. On MITI's influence on the petroleum industry, see Tanaka (1980).
27. See Ministry of International Trade and Industry (1969) for a detailed explanation of specific regulation measures in the low sulphurization plan.
28. Petroleum Association of Japan (1968).
29. *Shukan Enerugi to Kogai* (Energy and pollution weekly), 33, 17 October 1968.
30. Pollution prevention agreements are often treated in the same way as laws and rules, and are regarded as the 'third means of regulation' by the administration. Concerning the legal characteristics of pollution prevention agreements, see, for example, Harada (1972), pp. 153–74.
31. Many pollution prevention agreements had already appeared as means of regulation in the second half of the 1970s. However, pollution prevention agreements are regarded as arbitrary agreement formation processes between the parties concerned, and it is thought possible to regard this as a part of the dispute mediation system. On this discussion, see Terao (1993), p. 178, and also Harashina (1983). Matsuno and Ueta (1997) show the effectiveness of pollution prevention agreements as a means to implement air pollution prevention activities of major thermal power plants, by analysing the marginal cost of pollution prevention.
32. In some pollution prevention agreements between local governments and electric power companies, regulation of stack gas desulphurization facility installation was already seen as a part of the agreements. See, Denki Sangyo Shinbunsha (1972).
33. Petroleum Association of Japan (1971), pp. 175–8.

34. See Tanaka (1980). The operation ratio of heavy oil desulphurization facilities was 65.5 per cent in 1973, and fell to 54.0 per cent in 1974, according to Petroleum Association of Japan (1974), p. 8.
35. See Tanaka (1980).
36. Johnson (1982), p. 21.
37. Johnson (1982), p. 22.
38. See Coase (1960).
39. OECD (1994), p. 190.

References

Baba, Masao (1988) 'General Comments I', in Ryutaro Komiya, Masahiro Okuno and Kotaro Suzumura (eds) *Industrial Policy of Japan*, Tokyo, Academic Press Japan, 541–4.

Coase, Ronald (1960) 'The Problem of Social Cost', *Journal of Law and Economics*, 3, 1–44.

Denki Sangyo Shinbunsha (ed.) (1972) Kogai-boshi-kyotei *Soran: Denryoku-hen*, (Comprehensive List of Pollution Prevention Agreement) Tokyo, Denki Sangyo Shinbunsha (in Japanese).

Furuichi, Masatoshi (1979) 'Nihon ni okeru tei-iouka taisaku no rekishi', (History of Law Sulphur Policy in Japan), *Sangyo to kankyo: (Industry and Environment)* February (50–4) and March (75–81) (in Japanese).

Hara, Yonosuke (1992) *Ajia Keizairon no Kozu, (Structural Outline of Asian Economic Theory)* Tokyo, Riburopoto (in Japanese).

Harada, Naohiko (1972) *Kogai to Gyosei-ho, (Pollution and Administrative Law)* Tokyo, Kobundo (in Japanese).

Harashina, Sachihiko (1983) 'Amerika no kankyo-chotei: jumin tono rigai-chosei no atarashii apurochi', ('Environmental Mediation in USA: A New Approach to Conflict Resolution Among Residents') *Kankyo Joho Kagaku, (Environmental Information Science)* 12 (3), 45–50 (in Japanese).

Ikemoto, Yukio, Eiji Tajika and Yuji Yui (1984) 'On the Fiscal Incentives to Investment: The Case of Postwar Japan', *Developing Economies*, 22 (4), 372–95.

Imai, Ken'ichi (1976) *Gendai Sangyou Soshiki, (Modern Industrial Organization)* Tokyo, Iwanami Shoten (in Japanese).

Japan Development Bank (1976) *Nihon Kaihatsu Ginko 25 Nenshi, (25-Year History of Japan Development Bank)*, Tokyo, Japan Development Bank (in Japanese).

Japan Finance Corporation for Small and Medium Business (1984) *Chushokigyo Kinyu Koko 30 nenshi, (30-Year History of Japan Finance Corporation for Small and Medium Business)* Tokyo, Japan Finance Corporation for Small and Medium Business (in Japanese).

Johnson, Chalmers (1982) *MITI and the Japanese Miracle: The Growth of Industrial Policy, 1925–1975*, Stanford, CA, Stanford University Press.

Komiya, Ryutaro, Masahiro Okuno and Kotaro Suzumura (eds) (1988) *Industrial Policy of Japan*, Tokyo, Academic Press Japan.

Lee, Soo Cheol (2004) *Kankyo Hojokin no Riron to Jissai: Nikkan no Seido Bunseki wo Chusin ni, (Theory and Practice of Environmental Subsidy: Focusing on Institutional Analysis in Japan and Korea)* Nagoya, Nagoya University Press (in Japanese).

Matsuno, Yu and Kazuhiro Ueta (1997) 'Kokenho' (The SOx Rating of the Pollution-Related Health Damage Compensation Law), in Kazuhiro Ueta, Toshihiko Oka and Hidenori Niizawa (eds), *Kankyo Seisaku no Keizaigaku: Riron to Genjitsu,*

(*Economics of Environmental Policy: Theory and Reality*) Tokyo, Nihon Hyoronsha, 79–96 (in Japanese).

Ministry of International Trade and Industry (1969) *Tei-iouka Taisaku ni Tsuite*, (*Concerning Law Sulphur Policy*) Tokyo, Ministry of International Trade and Industry, October (in Japanese).

Miwa, Yoshiro (1988) 'Coordination within Industry: Output, Price and Investment', in R. Komija, M. Okuno, and K. Suzumura (eds), *Nihon no sangyou-seisaku* (*Industrial Policy of Japan*), Tokyo, Academic Press, 475–96.

Murai, Tadashi (1975) *Kogai Kazeiron, (Taxation Theory for Pollution*) Suita-city, Osaka, Institute of Economics and Political Studies, Kansai University (in Japanese).

Murakami, Yasusuke (1996) *An Anticlassical Political-Economic Analysis: a Vision for the Next Century*, trans. Kozo Yamamura, Stanford, CA, Stanford University Press.

Nakaoka, Tetsuro (1974) *Konbinato no Rodo to Shakai, (Labor and Society in Industrial Complex*) Tokyo, Heibonsha (in Japanese).

Niizawa, Hidenori (1997) 'Kankyo hojokin', (Environmental Subsidy) in Kazuhiro Ueta, Toshihiko Oka and Hidenori Niizawa (eds), *Kankyo Seisaku no Keizaigaku: Riron to Genjitsu, (Economics of Environmental Policy: Theory and Reality*) Tokyo, Nihon Hyoronsha, 191–202 (in Japanese).

O'Connor, David (1994) *Managing the Environment with Rapid Industrialisation: Lessons from the East Asian Experience*, Paris, OECD.

OECD (1977) *Environmental Policies in Japan*, Paris, OECD.

OECD (1994) *Environmental Performance Reviews: Japan*, Paris, OECD.

Ogura, Seiritsu and Naoyuki Yoshino (1988) 'The Tax System and the Financial Investment and Loan Program', in R. Komija, M. Okuno, and K. Suzumura (eds), '*Industrial Policy of Japan*', Tokyo, Academic Press Japan, 121–53 (in Japanese).

Oh, Sukpil (1999) *Kankyo Seisaku no Keizai Bunseki, (Economical Analysis of Environmental Economics*) Tokyo, Nihon Keizai Hyoronsha.

Okimoto, Daniel (1989) *Between MITI and the Market: Industrial Policy for High Technology*, Stanford, CA, Stanford University Press.

Petroleum Association of Japan (1968) *Sekiyu Gyokai no Suii, (Transition of Petroleum Industry) 1968 edition*, Tokyo, Petroleum Association of Japan (in Japanese).

Petroleum Association of Japan (1971) *Sekiyu Gyokai no Suii, (Transition of Petroleum Industry) 1971 edition*, Tokyo, Petroleum Association of Japan (in Japanese).

Petroleum Association of Japan (1974) *Sekiyu Gyokai no Suii, (Transition of Petroleum Industry) 1974 edition*, Tokyo, Petroleum Association of Japan (in Japanese).

Pollution Control Service Corporation (1991) *Kogai Boshi Jigyodan 25 nenshi, (25-Year History of Pollution Control Service Corporations*) Tokyo, Pollution Control Service Corporation (in Japanese).

Suehiro, Akira (1998) 'Hatten tojokoku no kaihatsu-shugi', (Developmentalism in Developing Countries) in Institute of Social Sciences, University of Tokyo (eds), *Kaihatsu-shugi (20 Seiki Sisutemu, Vol. 4.*), (Developmentalism: 20-Century System) Tokyo, University of Tokyo Press, 13–46 (in Japanese).

Suzumura, Kotaro and Masahiro Okuno (1993) 'Nihon no sangyou seisaku: tenbo to hyoka' (Japanese Industrial Policy: Perspective and Evaluation), in Noriyuki Itami, Tadao Kagono and Motoshige Itoh (eds), *Readings Nihon no Kigyou Sisutemu (Readings of Japanese Industrial System*), Tokyo, Yuhikaku, 146–82 (in Japanese).

Tanaka, Naoki (1980) 'Sangyo seisaku no minaoshi to gyosei kaikaku: Sekiyu gyoho ni miru gyosei kainyu no seika' (Review of Industrial Policy and Administrative Reform), *Keizai Hyoron (Economic Review*), November, 15–44 (in Japanese).

Teranishi, Shun'ichi (1993) 'Nihon no kogai mondai, kogai taisaku ni kansuru jakkan no shosatsu', (Pollution Problems in Japan: A Review of Pollution Control), in

Reeitsu Kojima and Shigeaki Fujisaki (eds), *Kaihatsu to Kankyo: Higashi Ajia no Keiken* (*Development and the Environment: The Experiences of East Asia*), Tokyo, Institute of Developing Economies, 225–51 (in Japanese).

Terao, Tadayoshi (1993) 'Taiwan: Sangyo kogai no seijikeizaigaku' (Taiwan: Policial Economy of Industrial Pollution), in Reeitsu Kojima and Shigeaki Fujisaki (eds), *Kaihatsu to Kankyo: Higashi Ajia no Keiken* (*Development and the Environment: The Experiences of East Asia*), Tokyo, Institute of Developing Economies, 139–99 (in Japanese).

Tsuruta, Kazumasa (1982) *Sengo Nihon no Sangyou Seisaku* (*Industrial Policy in Post War Japan*), Tokyo, Nippon Keizai Shinbunsha (in Japanese).

Tsuruta, Toshimasa (1988), 'Rapid Growth Era', in R. Komija, M. Okuno and K. Suzumura (eds), *Industrial Policy of Japan*, Tokyo, Academic Press, 49–87.

Ueta, Kazuhiro, Toshihiko Oka and Hidenori Niizawa (eds), (1997), *Kankyo Seisaku no Keizaigaku: Riron to Genjitsu* (Economics of Environmental Policy: Theory and Reality), Tokyo, Nihon Hyoronsha (in Japanese).

Wada, Yatsuka (1992) *Sozei Tokubetsu-sochi,* (*Special Measures for Taxation*), Tokyo, Yuhikaku (in Japanese).

Weidner, Helmut (1995) 'Reduction in SO_2 and NO_2 Emissions from Stationary Sources in Japan', in Martin Jänicke and Helmut Weidner (eds), *Successful Environmental Policy: An Introduction in Successful Environmental policy*, Berlin, Edition Sigma, 146–72.

Yamaguchi, Tsutomu (2002) *Chikyu Kankyo Saisei heno Chosen* (*A Challenge for Global Environmental Policy*), Tokyo, Jiji Press (in Japanese).

2
Historical Dynamic Interactions between Regulatory Policy and Pipe-end Technology Development in Japan: Case Studies of Developing Air Pollution Control Technology

Yoshifumi Fujii

The report entitled *Environmental Policies in Japan* by the Organization for Economic Cooperation and Development (OECD) reviewed Japanese environmental policy in the 1970s and its outcome, concluding that 'Japan has won many pollution abatement battles, but has not yet won the war for environmental quality' (OECD, 1977). The report characterized the policy as a non-economic or command and control one. The report also questioned how environmental improvement under the policy could be consistent with economic performance and industrial competitiveness. In fact, the Japanese economy achieved relatively good performance after succeeding in sharp reduction of the sulphur oxides (SOx) concentration and even after introducing the first emission standard for nitrogen oxides (NOx) in the world, which had the potential to cause more than a small economic impact.

Economic studies teach that stringent environmental regulation causes an increase in compliance cost for meeting the regulations and has a bad effect on competitiveness. However, there are other views, such as Michael Porter's argument, which is known as the 'Porter Hypothesis' (Porter, 1991), and which stresses the dynamic properties of the innovation triggered by strict regulation, saying that '[when] regulations are properly crafted and companies are attuned to the possibilities, then innovation to minimize and even offset the cost of compliance is likely in many circumstances' (Porter and vender Linde, 1995, pp. 97–118).

This chapter tries to answer this question of how Japan's environmental improvement could be consistent with economic performance and industrial competitiveness by drawing on the history and presenting empirical evidence. First, Sections 1 and 2 illustrate the dynamic interaction between the regulation and the development of control technology in Japan through case studies of sulphur oxides (SOx) and nitrogen oxides (NOx). The reduction amount per capita of SOx emission during 1965–70 attained the best ratio of any industrial country (Weidner, 1995), and the development of flue gas treatment technology accelerated progress in environmental policy in other countries, especially in Germany (Weidner, 1995). In NOx control, a very unique policy characterized as dynamic regulation in expectation of future development in abatement technology is analysed based on the empirical data.

Second, the 'success in 1970s' also left its mark on the Japanese economy and society. In Section 3, this study will discuss how Japan's race to the top in the 1960s and 1970s improved economic circumstances and, at the same time, brought about a fall into the non-flexible environmental policy of the 1980s.

Third, in Section 4, through both successful and stagnant experiences, this study will present some conclusions in response to the debate on the trade-off between regulation and competitiveness.

1.   Interaction between SOx control policy and technological development in Japan

The OECD report added two more points as major characteristics of the Japanese environmental policy of that day, 'concentrated control of specific pollutants and creation of a good spiral between regulatory policy and the development of pollution technology, which it called the policy of "Punish Polluters Principle (PPP)"'.

The Basic Law for Environmental Pollution Control (BLEPC), which was the first environmental law and represented an evolutional change, was approved in 1967; however, it still retained the traditional clause of 'considering harmonization between the economy and environmental regulation'. Facing environmental deterioration even after 1967, the Diet finally decided to delete the clause in 1970. This was the starting point of the successive steps of the reduction-first policy. The policy, oriented towards command and control and the development of pipe-end technology, is illustrated below.

Brief history of SOx regulation

The first policy aimed at effective SO_2 control in Japan was introduced in a local government, Yokohama City, in 1964. Facing higher SO_2 concentration (the peak average concentration in the urban areas in Japan was recorded in 1967), Yokohama City Government (YCG) succeeded in signing a contract on use of the low-sulphur coal with an electric power wholesaler, the public corporation Dengen Kaihatsu, when YCG approved the construction plan of a new coal-fired power plant. Without any available local government authority to control air pollution and faced with an ineffective national control law, YCG came up with the tactful idea of applying a different kind of agreement to the pollution control. This contract, which later came to be known as an 'agreement', was exchanged between the local government and the large-scale polluters and soon became a popular measure among local governments.

Fujikura (2002) argues that the measure was a form of administrative guidance provided by the local government rather than a voluntary agreement. At that time, YCG was also considering the possibility of having the power generator plants introduce flue gas desulphurization (FGD) equipment but finally gave up after concluding that introduction of FGD was technologically infeasible (Saruta, in Institute of Developing Economies, 2002).

In central government, there was no action worthy of special mention to control SOx until the middle of the 1960s. A new law (Smoke and Soot Regulation Law) to regulate smoke was enacted in 1962, but it had no actual effect on the pollution. Rather, it played towards killing the local government's activity in polluted areas by introducing national uniform standards (Fujikura, 2002). Under pressure from the local governments and expanded protest movements, anti-pollution sections were newly organized in the Ministry of Welfare (MW) in 1963 and in the Ministry of Trade and Industry (MITI) in 1964, in preparation for more drastic control measures at the national level. The establishment of these new anti-pollution sections was triggered by the terribly polluted and continually deteriorating condition in Yokkaichi City, where the peak concentration was recorded in 1964 (Hashimoto, in Institute of Developing Economies, 2002), and officials were also prompted to make preparations for the basic law by the environmental protest movement, which successfully halted the implementation of a big construction project involving a petrochemical complex in the Mishima, Numadu and Shimizu areas in 1964. An anti-pollution campaign by the media, a political battle in the Diet and the advance of the joint investigation of the polluted areas organized by both ministries impelled the government to enact BLEPC in 1967. Based on BLEPC, arrangements were made for comprehensive anti-pollution measures such as the Air Pollution Control Law (APCL) in 1968, the environmental quality standards for SOx in 1969 and the regional environmental pollution control programme in 1970.

APCL regulated the sulphur oxides concentration for each emission source depending on the seriousness of regional pollution and the height of chimneys based on the diffusion equation. Large-scale emission sources located in each region had to be equipped with tall chimneys of a height specified by the coefficient in an equation that was determined by the regional seriousness of the pollution. However, it became clear that only taking measures to promote diffusion could not satisfy the air quality standard; rather, this led to expansion of the polluted area. The next crucial issue was how to reduce the total sulphur content at the emission source.

In 1966, prior to BLEPC, through discussion at the Diet's industrial pollution session, it was already decided to let the government take measures promoting oil desulphurization (OD), starting in 1967, and FGD at the fuel-fired power station, starting in 1969 (Diet, industrial pollution session, 21 April 1966). The government initiated a plan to gradually increase the amount of the low-sulphur oil supply by subsidizing the construction of an OD plant starting in 1969. This original reduction plan decided by the Oil Council under MITI was later found to be insufficient to achieve the goal, but the oil refinery companies were against making additional investment to achieve the goal. So it took a time until the amendment plan was approved by stakeholders (Yamaguchi, in Institute of Developing Economies, 2002). In the earlier stage of SOx control, the focal point was how to reduce the emission from fuel-fired power stations, which were the largest polluters. Following the example of YCG, the Tokyo Metropolitan Government (TMG) exchanged an agreement with Tokyo Electric Power Company (TEPCO) to reduce SOx emission from its oil-fired power station in 1968. In this agreement, TEPCO promised to convert the fuel from heavy oil to crude oil, which has the lowest sulphur content. The negotiation was making slow progress, but TEPCO President Kikawada made the final decision on acceptance of this agreement (Kobayashi, in Institute of Developing Economies, 2002).

In the mid-1960s, Yokkaichi was the most polluted area, and it had become notorious for pollution-related asthma. In 1964, one acknowledged sufferer of 'Yokkaichi asthma' died. That caused SOx reduction in this area to become the major concern not only for the local government but also for the central government. The introduction of tall chimneys by some major polluters before enforcement of the APCL succeeded in reducing the peak concentration in the most polluted areas; however, taller chimneys led to the wider spread of the pollution. In 1972, five years after the initiation of legal proceedings, a judgement at the Yokkaichi branch office of the Tsu District Court ruled in favour of the complainant. This judgement had the remarkable impact of accelerating a flurry of SOx control policies nationwide. Soon after the judgement, the government approved laws and standards including an amendment to the ambient air quality standard (1973), the Pollution-related Health Damage Compensation Law (1974) and the Area-wide Total Air Pollutant Load Control Law (1974). The court judgement also had influence on the direction of control technology (see next section).

After the judgement of the Yokkaichi pollution lawsuit, the reduction-first policy at both central and local government levels remarkably reduced the SO_2 concentration, as seen in Figure 2.1.

Characteristics of the SOx reduction policy in the early stage

In the case of SO_2, reduction planning is easier than with other pollutants such as NOx or particulate matter because toxic material is discharged from nothing other than fuel contents and the emission source is limited to fixed-point sources in the industrial sector. When taking action aiming at reduction in SOx amounts, once the reduction goal is planned, all the control authority has to do is to allocate or ration the total amount of sulphur to be reduced in the fuel. The available reduction measures were limited in technological control; namely, they were OD for supplying low-sulphur oil to many small-scale users, FGD at large fuel-fired facilities and conversion to a fuel with lower sulphur content, such as low-sulphur crude oil and LNG.

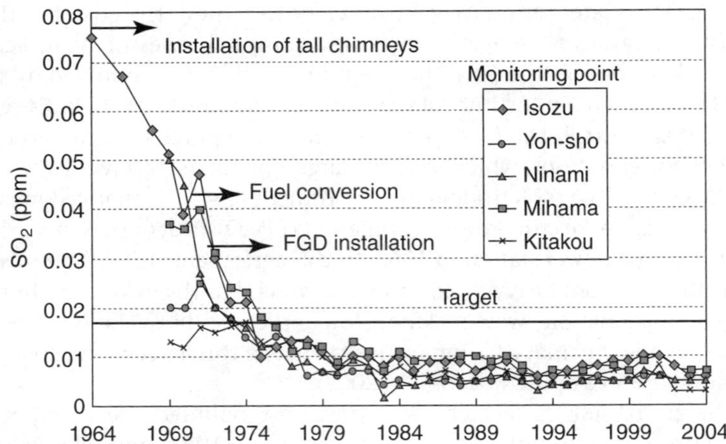

Figure 2.1 Change in SO_2 concentration at the monitoring points in Yokkaichi

Until the introduction of the 'reduction-first' policy was resolved by the Diet in 1970 and the birth of the Environmental Authority (EA) occurred, the issue was how to regulate the polluters and relieve the pollution victims' distress using conventional policy measures. The major factors in the success of the pollution battle in the early stage were: (1) strong protests by the public and the mass media against the pollution in light of precedents set by public nuisances such as the Minamata disease; (2) the small but significant level of discretionary powers for regulating industries given to the local government; (3) successful tactics by the authorities (MW) to protect public health, together with other authorities' (MITI's) efforts to regulate industry; and (4) the existence of available measures besides the environmental regulations to regulate industry, especially the public corporations.

Among the companies, the electric company played a great role in accepting the requests of the local government at an early stage, thereby breaking the deadlock situation between industry and the residents in the polluted area. The reasons for the electric company's acceptance can be explained as follow. First, the electric power company is regulated by MITI as a locally monopolistic corporation, and MITI has a variety of command and control measures in place for the electric company. Particularly important was the fact that the electric company had to apply for MITI's approval to alter the rates it charges. It is said that MITI approved higher electricity rates in exchange for the electric power company's introduction of FGD. Second, the electric company as a local monopoly acted to avoid conflict because it thought that conflict with the residents would have a negative effect on future construction plans for power plants. Also, in the NOx case, discussed in the next section, the local monopolistic scheme in Japan is a very important point when we explain the successful introduction of the strict regulation standards for the first time.

Such a scheme may have made it easier for the authorities, both those concerned with industrial policy and those concerned with environmental policy, to negotiate the possibility and feasibility of severer regulation through administrative guidance. In this context, lower transaction cost was one of the factors for explaining the feasibility of the reduction-first policy.

After the judgement of the Yokkaichi Court in 1972, the major concern shifted to how to reduce total emissions with feasible policy measures.

Development of flue gas desulphurization (FGD) technology

To achieve the ambient air quality standard in the polluted areas, it was clear that concentration-based regulation and supply of low-sulphur oil through OD were insufficient. Thereupon, the feasibility of installing FGD at large-scale combustion facilities became a focal point. The government had started an R&D project on FGD in 1966 as a subsidiary programme in MITI. Two technologies were selected and both of them adopted the process called the 'dry' process. Originally, there were two promising processes, the dry process and the wet process. In the dry process, no liquid is used and there is no need to reheat the gas. The advantage of the dry process was its cost and the desulphurization ratio it achieved, but it was at a disadvantage in terms of technological feasibility compared to the wet process. Although neither of the contracted technologies was ever commercialized, this project left the door open for discussion on whether such subsidized R&D programmes that assumed high risk were useful in the later remarkably successful commercialization of FGD technology in Japan (see Chapter 3).

Development of FGD technology has a long history. In 1850s, the first FGD process was proposed for use in sulphuric acid production, and the wet process FGD technology was developed in the 1930s, using alkaline Thames water in an effort to solve the London smog problem. At the starting point of the R&D programmes in Japan in the late 1960s, a technological menu based on development experiences in the USA and West Germany was already prepared. The critical R&D issues were in the selection of the dry process or wet process, in verifying the reliability by scaling-up the size, and in the cost reduction (Koizumi and Fujii, 1978).

The court judgement in Yokkaichi made clear the cause and effect in SOx pollution from an epidemiological standpoint and determined the polluter's monetary compensation based on the amount of discharged SOx emission. It was clear that the control measure with fuel conversion could not achieve the air quality standard in Yokkaichi, and this implied that the introduction of FGD was necessary as an additional measure. This not only increased the possibility of introducing FGD as a promising measure of the total emission control policy but also hastened the choice between the dry and wet processes. Despite its higher cost, the wet process was chosen as the most promising technology in terms of commercialization, which was the first goal.

In 'needs-pull' technologies, such as pollution technologies, which require immediate development, the lead time until their commercialization greatly depends on the time of convergence for choosing the most promising technology from among a variety of proposed technology systems (Fujii and Kikuchi, 1993). In this context, many of the engineering companies could make a decision without hesitation to adopt the wet desulphurization process. Also, they could forecast the total expected market size, on which they based their R&D expenditure. Before the court judgement, some large scale polluters in Yokkaichi started R&D programmes on the FGD technology. Chubu Electric Power Company (CEPCO) constructed a bench-scale R&D plant with the dry process in 1964 as a subsidiary to MITI's R&D programme mentioned above, and some other chemical and metal processing companies also started development of FGD technology in the early 1960s. However, once the regulatory scheme visibly shifted toward a more severe direction nationwide, the engineering companies rushed to find an opportunity to enter the technological competition. At that time, the dry process was abandoned, and the focus shifted to selection of the most feasible wet process. Around 1972, the lime-limestone wet process, the design of which was introduced from the company in the US as borrowed technology, was recognized to be the most promising method among the several wet process technologies.

Figure 2.2 shows the skyrocketing installation of FGD equipment after 1974, when the total emission control regulation was introduced in the polluted areas. The first pilot plant (dry process) was in operation in 1968,

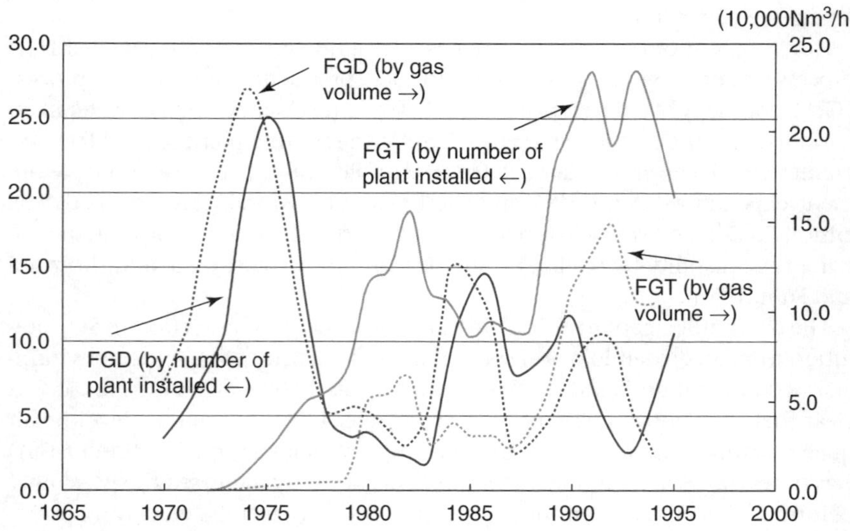

Figure 2.2   Trends in the annual commercialized capacity of FGD and FGT installed in Japan (data represents the moving average)
Source: Adopted from the website of the Environmental Restoration and Conservation Agency of Japan, http://www.erca.go.jp/english/index.html

five years after the launch of R&D in 1963, and the first commercialized plant was installed in 1972 (and went into operation in 1973), nine years after the launch of R&D. As shown in Figure 2.2, the FGD market peaked at an early stage, soon after the total emission regulation was introduced in 1974, and the market faced stagnation after 1978. The average yearly market scale of OD and FGD during 1981–99 amounted to about $0.2 billion and $0.4 billion, respectively. Such stagnation was foreseeable and FGD manufacturers appeared to be seeking the next market.

2.   Case study in NOx control policy and the development of flue gas treatment (FGT) technology

The dynamic and mutual interaction between regulatory policy and the development of the NOx technology in Japan was very interesting. In particular, a point that differs from the SOx case is the answer to the question of why the competition in the 'race to the top' was emerging among widespread industries without any large, subsidized R&D programme. The EA, just after its foundation in 1971, played the leading part in this race. As in similar cases in other countries, the EA was sounding out the possibility of establishing emission standards with the support of society in the face of terrible pollution. However, the result in the NOx case in Japan was quite different from that in the US. Besides social, political and legislative factors, interaction between policy and technological development and industrial organization are the keys to explaining this difference.

Policy background of NOx regulation in Japan

Although the NOx pollution was not yet widely recognized in Japan and its control was believed to be harder than SOx control, Japan introduced the first emission standard in the world and successfully developed the control technologies in a short time. Both the mechanisms of NOx formation and of the transition from NOx to the direct pollutants are more complex. In NOx pollution, not only nitrogen content in the fuel but also the peak temperature, which differs depending on the combustion equipment, affect the amounts of NOx emission; therefore, emission standards vary depending on the combustion technology, unlike in the SOx case. Furthermore, the emissions from automobile engines, which have a higher combustion temperature, had to be taken into consideration. In addition to the toxicity of NOx itself, it forms oxidants that cause photochemical smog through very complex and unstable chemical reactions, and it also forms particulate matter (PM); therefore, it was very hard to introduce a comprehensive policy to control the NOx-related pollution.

When we talk about the interaction between regulation and technology in Japan, we cannot ignore the US policy that forms the background of Japanese policy. Following the terrible smog caused by NOx emission in Los

Angeles in 1943, basic NOx-related information became available; developments included presentation of the formation mechanism of the Los Angeles smog in 1952, the control technology such as low NOx combustion technology for the point source in 1959, the first automotive emissions control technology in 1961 and an organization for control policy in the State of California in 1947 (Koizumi and Fujii, 1978). Since 1967, when this type of the pollution was widely seen in US, the US federal government began to take leadership in NOx control policy and the National Air Pollution Control Administration (NAPCA), a former body of the US Environmental Protection Agency (EPA), implemented surveys and R&D programmes on a larger scale. By 1970, a considerable technological menu to control NOx was already prepared. In the US, the focal point was the regulation of automobile emission, which was a major NOx polluter. US Senator Edmund Muskie was instrumental in the adoption of a dynamic and 'technology-forcing' policy in the Clean Air Act of 1970, which was amended to mandate that the automotive manufacturers commercially market an engine that removes 90 per cent of NOx by 1976. But the Act was abandoned four years after its approval in the face of opposition from the automotive industry. Although the policy of dynamic and technology-forcing improvement was not realized in the US, the dynamic regulation later realized in Japan was originally seen in the Clean Air Act of 1970.

Just when SOx concentration was drastically improving in Japan, a midsummer incident that was likely caused by photochemical smog afflicted students at Rissho High School during an outdoor class in 1970. The Air Pollution Authority of the TMG, supported by public opinion against pollution, took immediate action on this new issue. In the same year as the incident, the TMG concluded it was caused by photochemical smog from NOx emission and soon set up a survey to collect emission data from the source and start monitoring. The EA also soon prepared a monitoring system and arranged a study on NOx control measures.

Technological development of NOx control in Japan

Non-point source (automobile)

In NOx control, the emission from automobiles was the centre of attention. At the end of the 1960s, the Japanese automotive industry was on the verge of expanding its exports to the US, and the Clean Air Act of 1970 proposed in the US by Senator Muskie was thought to be a kind of menace for promotion of Japanese exports to the US. Before there was any reaction in Japan, Japanese auto-makers were considering doing research and testing to refute the Muskie Act, which required a 90 per cent NOx reduction by 1976. However, after the EA decided to introduce an automotive emission standard in Japan by just reprinting the Muskie regulation, auto-makers had to meet simultaneously the regulation for both the US and Japan. The fact that the Automobile Expert Committee in the EA had to adopt the Muskie regulation in the absence of an environmental standard in Japan shows the seriousness of the air pollution and how the EA was forced to quickly prepare a countermeasure against it, without even considering not only the economy and but also the scientific basis of NOx emission and its impact mechanism. In Japan, the government decided in 1970 to introduce a loose NOx emission standard that applied to 1973 model year cars and decided in 1972 to strengthen the standard for 1975 model year cars. After a short time, in 1973, the EA put the national air quality standard for NO_2 into legislation based on epidemiological survey data, which later resulted in a big dispute over its scientific basis in the Diet. At last it was deregulated in 1978.

The year 1974 was an epoch-making year in the history of NOx control in Japan. One committee in the EA held a hearing with the auto-makers about the technological feasibility of achieving the Muskie Act. The Anti-pollution Authority of the TMG also discussed the technological feasibility, and its final report concluded that 'the 1976 regulation is feasible'. During the battles among the EA, local authorities and the auto-makers, two auto-makers suggested that it may be possible to achieve the 1976 standard. These companies were Toyo Kogyo (Mazda) and Honda, both of which were classified as small- or medium-scale companies, and they succeeded in developing low-NOx engines in a different way (non-catalytic technology) from the other companies. Nishimura (1976), a committee member in the TMG, explained their behaviour thus: 'The stricter standard was a good opportunity for both automakers to expand their market share because they had technological advantages in the competition to develop the low-NOx engine.' Finally, the committee at the EA decided to rescind the 1976 regulation shown in the Muskie Act and to postpone the regulation until 1978. In its place, the committee decided to introduce a provisional standard enacted in 1976 that was a little bit looser. At any rate, Japan decided to introduce the automotive NOx emission standard, while the Muskie Act was abandoned in the US after a political battle between the environmental protectionists and the automotive industry in 1974. Incentive to increase market share by ignoring the stricter regulation had broken the deadlock. A different engine equipped with a three-way catalyst that finally met the 1976 regulation came to be the mainstream technology for meeting the 1978 regulation. This type of electronically controlled engine remarkably improved automotive fuel economy. All Japanese auto-makers successfully developed this type of engine, and this enabled them to expand their exports, especially after the second oil crisis in 1979.

Fixed-point source (stationary combustion source)

In response to the regulatory reaction by the central and local authorities mentioned above, many R&D programmes were established. The first research on flue gas treatment was undertaken as basic research at the national laboratory under MITI in 1969. The major concern of the private sector for the low-NOx technologies in early stage was in developing combustion technologies such as low-NOx burners and lean-burn systems with cost advantages.

However, predictable growth in the FGD market, which peaked in 1975, drove nearly 50 private companies to establish FGT research during 1971–5. It can be said that Japanese administrators and experts had obtained almost all of their information through journals and reports published by APCD of Los Angeles County and NAPCA (EPA later). For example, the Bartok Report (1969), published by NAPCA, was the comprehensive technology assessment report on FGT, and it was widely read among researchers and engineers in Japan. As in the SOx case, FGT equipment was not yet commercially marketed anywhere in the world, but a rough selection of promising processes was already reviewed, based on the US experiences mentioned above.

As in the SOx case, at the time when the EA started the study for NOx control, the focus in FGT development was to select the most promising technology. Selection of the dry process or wet process and the selection of a reliable catalyst (in the case of the dry process) and absorption solution (in the case of the wet process) were the major theme. Because the wet process required reheating of the flue gas, resulting in high cost, the more feasible dry process that did not require reheating was believed preferable. In the early stage, ammonia with or without a catalyst was selected as the most feasible NOx removal agent in the dry process. In the dry process, more than 80 per cent of NOx was removed with a catalyst and 40–50 per cent without a catalyst. The wet process costs more than the dry one, but it was designed for use on emission sources with higher NOx concentration.

The process of selecting promising technology in FGT is illustrated in Table 2.1 in terms of lead-time. Data in parenthesis shows the number of R&D plants tested. The table shows that the dry process with a catalyst won

Table 2.1   Lead-time and commencement time of FGT project: months (number of plants)

Process alternatives	Commencement	Lead-time (R&D–Bench)	Lead-time (Bench–Pilot)	Lead-time (Pilot–Completion)
Dry ammonia and catalyst				
for clean gas	1971–4	9.6 (8)	10.7 (6)	28.2 (5)
for semi-dirty gas	1971–4	9.5 (8)	13.8 (8)	43.5 (6)
Dry ammonia and catalyst				
for dirty gas	1973	–	13.5 (2)	59 (1)
Dry ammonia and non-catalyst process	1972–5	0 (3)	12 (1)	20 (1)
The other dry process	1971–2	8 (2)	7 (2)	–
Wet process	1971–4	6.2 (8)	8.5 (10)	19 (2)

Source: Fujii, 1978.

the race and that the dirtiness of the flue gas affected the lead-time in the dry process. In the FGT case, it took at a minimum of four years to be commercialized (Koizumi and Fujii, 1978).

The EA introduced the first NOx emission standard for large-scale boilers and sources emitting higher concentration NOx gas in 1973, but it was a temporary, loose standard. During only six years from 1973–9, the EA introduced a standard for stationary emission sources step-by-step, by tightening the standard and by extending the regulation target four times. The favourable outcome of the R&D in the 'race to the top' suggested to the EA that it was possible to adopt the best available technology. Up to and including the third standard in 1977, the policy of implementing stricter regulation by making use of technological developments seemed to be successful. Based on the technological review in 1977, the EA decided to introduce the fourth standard, which applied not only to the polluted areas but also nationwide, in 1979.

In the end, however, the fourth emission standard did not require the installation of FGT, which very much disappointed the FGT developers who were expecting a new regulated market. The race to the top was slowing down at the end of the 1970s in the face of a rally by the industries. The Japan Iron and Steel Federation, which was the most influential among the stakeholders, started a campaign against the unscientific air quality standard for NO_2. An iron and steel company installed a large-scale FGT plant for a sintering furnace based on an agreement with the local government, but the Federation was afraid that further installation of such expensive equipment would impair companies' ability to compete. At last, the standard was relaxed after a debate during the Diet session on 27 July 1978. As the result, the fourth NOx emission standard was implemented without introducing FGT except at limited facilities (i.e., large-scale emission sources with clean gas). This resulted not only in greater shrinkage of the market scale than expected but also in exclusion of the smaller FGT developers from the market because large boiler owners purchased the same brand of FGT plant. One catalyst manufacturer that played a very important role in making the development of dirty gas treatment more feasible went bankrupt because the emission sources with dirty gas were not included in the regulation target. As shown in Figure 2.1, the market for FGT was smaller than that for FGD before 1990 and far smaller in the early stages. For the firms that entered into the race, the regulation ended up being such that only the big names skimmed the cream in the 1980s.

3.   After the race to the top

Although the Japanese experience with winning the pollution abatement battle in the SOx and NOx cases in a short time was very remarkable, the short-term battle was accompanied by impact not only on environmental policy but also on the economy and society.

H. Weidner (1995) points out that the successful outcome was achieved:

by the complex combination of various flexible managed regulatory policy instruments, the use of 'meta-instruments,' political pragmatism, the capability of reaching consensus among strategic groups with respect to environmental targets, tremendous political pressure on the government and industry from environmental movements as well as innovative behavior by local administration.

Adding to these factors, the interaction between the reduction-first policy and the technological development in the private sector cannot be ignored. In fact, dynamic regulatory policy in anticipation of technological development aiming towards an environmental target, as well as the technological race among private companies in anticipation of a new regulatory market were the key factors, especially throughout the race to the top.

'Innovation offset' after the oil crisis that connects to industrial modernization

Spillover effects in the wider economy could be seen after the race to the top. During the race mentioned above, the first oil crisis in 1973 resulted in skyrocketing oil prices, but it caused a tailwind for energy-intensive industries such as the iron and steel industry, the chemical industry and the machinery industries producing energy-saving products. Auto-makers and many other manufacturers succeeded in expanding their exports, especially after the second oil crisis in 1979. This was the outcome of introducing the reduction-first policy. Automobile emission control technology was connected to the lean-burn (fuel-efficient) engine with electronic controls; the low-NOx burner resulted in higher energy efficiency; and the introduction of FGD at the large fuel-fired facilities promoted heat recirculation. These three factors combined to produce a competitive advantage for the products.

Thus, economic adjustments induced by the environmental regulation prior to the energy crisis fortunately were synchronized with the changes that were later induced by higher energy prices. As a result, the Japanese economy was able achieve a relatively good performance compared to the other countries shown in Figure 2.3. This was an unintentional result; the regulatory authorities never planned such a dynamic spillover effect in advance. In fact, in the arguments for and against introducing NOx regulation between MITI and the EA, the economic impact of the NOx regulation using econometric models was discussed. The EA insisted that little impact would occur because enhancement of the investment in environmental equipment could compensate for the negative impact, while MITI forecasted a significant negative impact on future economic growth due to cost increases. But in this debate, the dynamic path that actually occurred later was never discussed.

Japan's achievement of economic development with nearly zero (4 per cent) increase in energy demand from 1973–86 and with far less pollution for a long time was realized through technological developments (Fujii *et al.*,

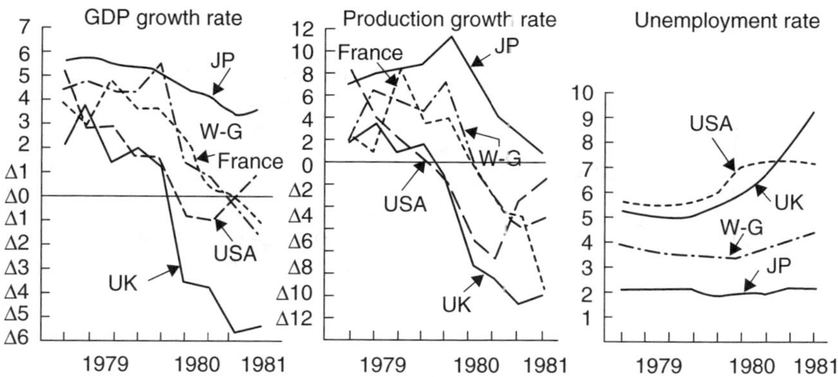

Figure 2.3   Economic performance following second oil crisis
Source: Economic Planning Agency, White Paper 1982.

1992; Fujii and Matsukawa, 1993), and this was evidence supporting the idea of 'ecological modernization', which has become a mainstream faction in Europe (Fisher and Freudenberg, 2001; Weidner, 2002). This idea has also led to the construction of an alternative concept for the next generation industrial society, including possible environmental improvements that can be achieved through technical innovations that go beyond end-of-pipe approaches, and this has opened up both economic and political opportunities under the condition of a maximum of market penetration (in material consumption) (Jänicke *et al.*, 2004).

Non-flexible structure of environmental protection

However, the reduction-first policy accompanied by command and control and subsequent scaling back of the environmental policy appears to be creating a kind of distortion or non-flexible structure in both the economy and society.

Economic dimension

Matsuno (Matsuno and Ueta, 1997) estimates the marginal SOx abatement cost in 1975 at Sakai Power Station in Osaka. ● in Figure 2.4 shows the actual cost based on the agreement between Sakai City and Kansai Electric Power Company (KEPCO). He compares the cost with the SOx rating introduced by the Air Pollution Authority of Osaka City Government prescribed by the Pollution-related Health Damage Compensation Law and points out that the measure actually introduced has a far higher cost than the regulation. The policy of 'reduction first' was very effective for abating pollution during 'the battle' (OECD, 1997) at the beginning stage; however, it also formed a higher cost structure for environmental protection and dampened the possibility of taking more flexible control measures in keeping with the energy price changes

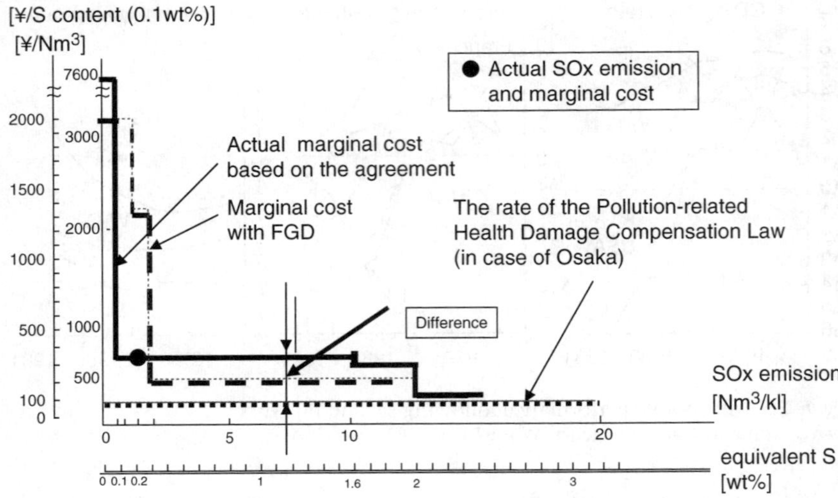

Figure 2.4   SOx rating of the Pollution-related Health Damage Compensation Law
and actual marginal cost at the Sakai Power Station
Source: Matsuno and Ueta, 1997.

afterward, and in fact the command and control policy is believed to be non-flexible in general.

In the world market for environmental products, Japanese environmental equipment manufacturers and service suppliers, except for those in the air quality and water quality market, receive a low evaluation (Berg and Ferrier, 1988). Since the 1980s, in many segments of the environmental field, Japan seems to be locked in a very heterogeneous environmental market from the other countries, characterized by very expensive and decorative technology. This may be a result of the fact that the Japanese environmental equipment industry has tended to neglect the foreign market mainly because the domestic market has paid well, being protected by the regulation.

Social dimension

Even now Japan is called a bureaucrat-centred society. Also in pollution issues, although several large local governments located in urban areas thought out a variety of unique policies, like anti-pollution agreements to control air pollution during the 1960s in the absence of national regulations, the locus of policy shifted to the central government after BLEPC in1967. In the central governmental organization, 100 years has been spent forming a peculiar power structure with vertical administrative features. Even after the Environmental Agency was founded in 1970, some ministries that were formerly responsible for certain industries still retained their power in environmental policy over those industries. One typical example is the case of the Environmental Impact Assessment

(EIA) Bill. Although an EIA draft was prepared in Japan soon after EIA approval in the US in 1969, it actually took 26 years to pass the EIA Bill, which was rejected three times after it was introduced at the national level. The strange thing is that EIA had been actually implemented until 1996 through the ministries. EIA for power stations was conducted by MITI, and EIA for the highway construction was done by the Ministry of Construction, and so on. These EIAs did not fully include a rational decision-making process with civil participation and disclosure of the administrative information. After the 1980s in Japan, people believed the myth that environmental issues were almost conquered, and people boasted about the Japanese experience in the 1970s. In such circumstances, the centralized environmental administration with a form of decentralized power balance among the ministries was maintained, while the environmental issues that could not be easily resolved without a more sophisticated scheme, such as environmental governance, occupied a prominent position in environmental issues. It was in the 1990s when the Administrative Procedure Law and the Law Concerning Access to Information held by Administrative Organs, both of which are essential to environmental solutions nowadays, were approved for the first time in Japan.

Socially, an active environmental protest movement that faced the terribly polluted situation supported the realization of the 'reduction-first' policy; however, the movement could not transform itself into a well-organized environmentalist group afterwards in Japan, while the same kind of protest movement grew up to be an international environmental NGO or create a single-issue political party in European countries. The worldwide environmental NGOs such as Greenpeace and Friends of the Earth were established around this time (1971), and in West Germany, a procedure type of law system was introduced in 1976. Due to the improvement in visible pollution and the stagnation of environmental policy, these movements were diminishing until recent years. Since the pollution problems were mainly caused by the factories, the environmental issues shifted to the more complex ones where the cause and effect of the pollution was not clearly distinguished and the ones where regulatory policy is not necessarily effective. In the 1970s, the success of the environmental policy fully depended on regulatory and technological solutions, and the conventional regime was locked into this mindset, which prevented it from reorganizing its environmental governance (Terao, 2002).

4.   Interaction between policy and technology in the context of the debate on Porter's hypothesis

Finally, it may be meaningful to examine how the above-mentioned interaction between the reduction-first policy and the technological development of pollution control could answer the debate in the context of the 'Porter Hypothesis'. This hypothesis was triggered by the discussion of Michael Porter (Porter, 1991), who insists that stringent environmental regulation can promote competitiveness through a more dynamic path, which brings benefits to offset the static loss, but another view from economists claims that the Porter Hypothesis cannot easily occur either theoretically or empirically. Since the 1990s, a considerable number of papers has been published on this debate (Jaffe, 1995; Jenkins, 1998).

Oates, Palmer and Portney (1993) performed a comprehensive examination of the possibilities of the Porter Hypothesis from a theoretical economics viewpoint. They showed that an increase in stringent regulation worsens sales and profits despite the adoption of more efficient technology when using a simple general equilibrium model, except in the case where the some important elements are missing in the model. The elements are:

(1) strategic behavior between pollution firms and between these firms and the regulatory agency, (2) the presence of an industry that produces abatement technology and equipment as its output, and (3) the existence of opportunities for profitable innovation in the production of the firm's output for some reason have been overlooked and somehow become realized in the wake of new and tougher environmental regulations.

The study also goes on to examine these elements; however, the extension studies of these elements are not fully incorporated into their model. Regarding the first element, an example is the case in which the government can actually improve the international competitive position of domestic exporters by imposing environmental standards upon them. The third element requires theoretical game analysis, and the second element requires insights from the industrial organization.

Regarding the empirical analysis to test the relationship between competitiveness and the stringent environmental policy, Jenkins concludes that there was no strong universal relationship between them, either at the firm level or the industry level, mainly because environmental regulation is only one of number of factors that may have an impact on competitiveness. Also, he mentions that '[neither] the conventional wisdom (economics) nor the revisionist view (supporting the Porter Hypothesis) receives unambiguous support from existing evidence' (Jenkins, 1998).

It seems very difficult to explain precisely the relationship between competitiveness and stringent regulation. In this paper, the competitive advantage structure in Japan was illustrated soon after the second oil crisis; however, this structure could not be adapted to the lower energy price after 1987. Since the end of 1980s, the Japanese economy went into stagnation. However, the stagnation can be explained by many other reasons, and the relationship between change in competitiveness and the regulatory policy is not easily proved. Rather, it may be more valuable for further studies to show some important condition or element that creates a dynamic offset process between regulation and technological progress. In this study, some evidence seen in the experience of the 1960s and 1970s, which is relevant to the debate, will be discussed below.

The first point is why it was possible to accelerate technological development in Japan, while the same kind of policy was abolished in US. Looking at case studies on the Japanese experience in the 1960s and 1970s, regarding both automobile and stationary equipment regulation, there was an expectation in the private sector for future market creation through new regulations, the implementation of which was contingent on their own success in technology; this created a strong incentive which led them to 'rent seeking' activity. However, at the same time, it should be noted that such a dynamic process was never created with a monopolized market and without an innovator who exhibited rent-seeking behaviour in the engineering industry. In fact, the oligopoly structure in the US automotive industry never realized the stringent regulation. As Oates *et al.* pointed out, the presence of an industry that produces abatement technology and equipment as its output is also important. Thirty-three firms out of 55 that entered into the FGT development race were engineering companies, which did not need to pay the compliance cost for the stringent regulations as polluters. In general, economic theory assumes that it is the polluting firm that faces the technological choice under regulations. However, the important point to note in the dynamic process mentioned above is the role played by the engineering companies as rent seekers, wherein they found a great incentive to create a large, regulated market through technological success. In this context, more discussion from the market structure viewpoint will be required. Figure 2.5 illustrates the dynamic process of developing FGT technology. It shows that, starting in 1969, many private companies established R&D projects, seeking to enter into the unseen market, and proposed a variety of technological processes. Technologies with commercial promise were selected and converged after the

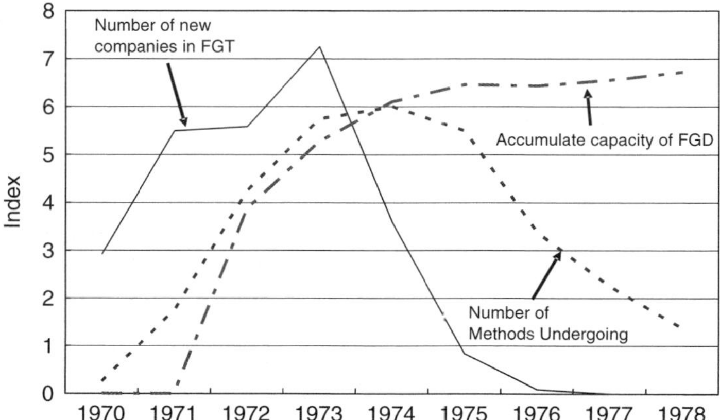

Figure 2.5   Dynamic process of developing FGT technology
Source: Fujii, 1978.

peak of new companies entering the market around 1973, and commercial plants were increasingly installed.

In the case of automobiles, two of them stole a march to protect themselves from the stricter regulation. This is a textbook case of the game theory (e.g., Ito *et al.*, 1988) from economics wherein a smaller company sometimes has an incentive for R&D expenditure that exceeds a socially appropriate expenditure level when it faces competition in a situation where it has nothing to lose. In fact, smaller companies had great incentives for stricter regulation because stricter regulation presented a good opportunity for them to extend their market share.

R&D policies differed in the cases of SOx and NOx. In the SOx case, the court judgement noting the need for additional control measures using FGD played a very important role. SOx R&D policy was a normal technology-push type policy that received subsidization in its early stage. In the NOx case, the R&D policy was a very unique technology-pull (purchase of feasible technology) one, where the authorities only suggested the possibility of stringent emission standards in the future. The idea of dynamic regulation was originally invented in the US (Senator Muskie), but the control authorities in Japan applied dynamic regulation without subsidies for R&D on stationary sources.

Second, why could an innovation offset be realized after the oil crisis? The answer at which this study arrived is that a dynamic movement was created where pollution control technology could be converted into energy conservation technology. It happened that the attention directed towards the 'reduction-first' policy was transformed into attention directed towards the energy crisis. If the energy price had been decreasing at the end of the 1970s, what would have happened to the Japanese economy? The idea of regulation designed to promote innovation or regulatory coordination looks like a very difficult task for the regulators. Indeed, environmental policy should be synchronized with industrial policy to ensure the outcome, but who can integrate the regulatory policy by forecasting for the future?

5.   Implications for future issues

Recently, in the issue of global warming, we frequently see an analogous discussion on the relationship between regulation and technology or competitiveness. On 28 March 2001, the President of the US, George W. Bush, announced that the United States would not implement the Kyoto Protocol on global warming, and his administration asserted that mandatory greenhouse gas reductions would be prohibitively expensive, cutting millions of jobs, costing gross domestic product and harming US competitiveness. European countries and Japan are entering into the Kyoto Protocol and are just starting to reduce CO_2 to attain the target. Under the Kyoto Protocol, a variety of control measures including economic instruments are to be introduced; a very unique regulatory scheme, the Top Runner Regulation (Ministry of Economy, Trade and Industry), which imposes an efficiency cap on the major electric appliances, was seen to control the CO_2 emission in Japan. Also for controlling CO_2 emission, a hybrid regulation that combines a carbon tax with a subsidy for the most promising facilities to reduce the CO_2 emission (National Laboratory of Ministry of Environment) was proposed. If the expectation is that energy prices will be higher for a long time hereafter, this study concludes that the first-mover country to adopt a more stringent regulatory policy on carbon dioxide will create an innovation offset and will briefly enjoy a competitive advantage as a nation.

References

Bartok, W. (ed.) (1969) *Technology Assessment of the Flue Gas Treatment Technology*, NAPCA.

Berg, David R. and Grant Ferrier (1998) *The U.S. Environmental Industry*, Office of Technology Policy, US Department of Commerce.

Environmental Agency (1972–80) White Paper.

Fisher, D. R. and W. R. Freudenberg (2001) 'Ecological Modernization and Its Critics: Assessing the Past and Looking Toward the Future', *Society and Natural Resources*, 14; 701–9.

Fujii, Yoshifumi (1978) 'Chisso-sankabutsu boushi-gijutsu nikansuru kosatsu' (A Study on Technological Development of NOx Control) (B.A. graduation thesis), Tokyo, Graduate School of Science and Technology, Waseda University (in Japanese).

Fujii, Yoshifumi Isao Matsukawa, and Seishi Madono. (1992) 'Chikyu-kankyo-mondai ni keizaiteki-syudan ha yukoka' (Are economic instruments effective on the global warming issue?) *Journal of Economic Seminar* 449; 12–19 (in Japanese).

Fujii, Yoshifumi and Junichi Kikuchi. (1993) *Sentan-gijutsu to keizai (High technology and economy)*. Tokyo; Iwanami Shoten (in Japanese).

Fujii, Yoshifumi and Isao Matsukawa (1993) 'Price, Environmental Regulation, and Fuel Demand: Econometric Estimates from Japanese Manufacturing Industries.' *Energy Journal*, 14 (4).

Fujikura, Ryo (2002) 'Nihon no chihou-kokyodantai no Iou-sankabutsu-taisaku: kodo-keizai-seichoki ni jissisareta kogai-boshi-kyotei to gyosei-shido' (Local Measures in Japan for Controlling Sulfur Oxide Emissions: Pollution Control Agreements and Administrative Guidance Adopted during the Era of Rapid Economic Development). In *'Kaihatsu to kankyo' no seisaku-katei to dainamizumu: nihon no keiken, higashi-ajia no kadai (Dynamism of the Policy Process in 'Development and the Environment': The Experiences of Japan and Problems in East Asia)*. T. Terao, and K. Otsuka, (eds) Chiba: Institute of Developing Economies (in Japanese).

Hamamoto, Mitsutsugu and Kazuhiro Ueta (1996) 'Kankyo-kisei to gijutsu-kakushin' (Environmental Regulation and Technological Development). In Ueta, K., M. Ishikawa, S. Okuda, H. Deguchi, M. Hamamoto, S. Fujisaki and E. Hosoda. *Atarashii sangyo-gijutsu to shakai-shisutemu (The new industrial society and its social system)*. Tokyo; Nikka-Giren (in Japanese).

Hanley, Nick, Jason P. Shogren and Ben White (1997) *Environmental Economics in Theory and Practice*. Macmillan Press.

Institute of Developing Economies (2002) Nihon no kogai-taisaku-keiken nikansuru hiaringu no kiroku (Interview record on the anti-pollution policy experience in Japan.) Chiba, Institute of Developing Economies (in Japanese).

Ito, Motoshige, Kazuharu Kiyono, Masahiro Okuno, and Kotaro Suzumura (1988) *Snagyo-seisaku no keizai-bunseki (Economic analysis of the industrial policy)*. Tokyo, Tokyo University Press (in Japanese).

Jaffe, A. with S. Peterson, P. Portney, and R. Stavins (1995) 'Environmental Regulation and the Competitiveness of U.S. Manufacturing: What Does the Evidence Tell Us?', Alan M. Rugman and John J. Kirton (eds), *Trade and the Environmental, Legal and Policy Perspectives*, Edward Elgar.

Jänicke M. and H. Weidner (1995) *Successful Environmental Policy: An Introduction in Successful Environmental Policy*, Berlin, Edition Sigma.

Jänicke Martin, (2004) 'Industrial Transformation between Ecological Modernization and Structural Change', in *Governance for Industrial Transformation*, from proceedings of the 2003 Berlin Conference on the Human Dimension of Global Environmental Change, ed. Klaus Jacob, Manfred Binder and Anna Wieczorek, Berlin, Environmental Policy Research Centre, 201–7.

Jenkins R. (1998) 'Environmental Regulation and International Competitiveness: A Review of Literature and Some European Evidence', Discussion Paper #9801, United Nations University.

Koizumi, Mutsuo and Y. Fujii (1978) 'Nenshou-souchi karano NOx hassei-boushi-gijutsu no kaihatsu nikansuru shiryou' (Data on development of the NOx control technology (I).). *Journal of Fuel Association* 57(69) 43–60 (in Japanese).

Matsuno, Yu and Kazuhiro Ueta (1997) 'Kouken-hou fukakin' (The SOx rating of the pollution-related health damage compensation law). In *Kankyo-seisaku no keizaigaku: ririon to genjitsu (Economics of the Environmental Policy and Its Interrelation with Society)*. Edited by Ueta, K. Tokyo, Nihonhyoronsha (in Japanese).

Nishimura Hajime (1976) *Sabakareru jidousha (Automotive industry judged in Japanese.)* Tokyo; Chuo-koron (in Japanese).

Oates, W. E., K. Palmer and P. R. Portney (1993) 'Environmental Regulation and International Competitiveness: Thinking about the Porter Hypothesis', Discussion Paper 94–02, Resource for the Future.

OECD (1977) *Environmental Policies in Japan*, Paris, OECD.

Palmer, K. and W. E. Oates (1995) 'Tightening Environmental Standard: The Benefit-cost or Non-cost Paradigm?', *Journal of Economic Perspectives*, 9(4), 119–32.

Porter, M. E. (1991) 'American's Green Strategy', *Scientific American*, (April) 168.

Porter, M. E. and C. vender Linde (1995) 'Toward a New Conception of Enviro-nment–Competitiveness Relationship.' *Journal of Economic Perspectives*, 984, 97–118.

Terao, Tadayoshi (2002) 'Kaihatsu to kankyo' no seiji-keizaigaku wo megutte: seisaku to shakaihendou' (Political Economy of 'Development and the Environment': An Introductory Analysis of Policy and Social Transformation) in *'Kaihatsu to kankyo' no seisaku-katei to dainamizumu: nihon no keiken, higashi-ajia no kadai (Dynamism of Policy Process in 'Development and Environment: The Experiences of Japan and Problems in East Asia)*. Edited by T. Terao, and K. Otsuka, Chiba, Institute of Developing Economies (in Japanese).

Weidner, Helmut (1995) 'Reduction in SO_2 and NO_2 Emissions from Stationary Sources in Japan', in M. Jänicke and H. Weidner (eds), *Successful Environmental Policy: An Introduction in successful Environmental Policy*, Berlin, Edition Sigma, 146–72.

Weidner, Helmut (2002) 'Capacity Building for Ecological Modernization Lessons From Cross-National Research', *American Behavioral Scientist* 45(9) (May), 1340–68, Sage Publications.

3

Role of Local Governmental Research Institutes in Development and Diffusion of Pollution Control Technologies in Japan

Yasushi Ito

Introduction

When certain kinds of environmental regulation are introduced, enterprises are, in many cases, forced to attain the target technologically. The types of environmental policy instruments that exert a strong incentive to develop or diffuse environmental technologies have been discussed theoretically in environmental economics.[1] Most of the analysis, however, regards the response of enterprises to environmental policies as a 'black box', that is, enterprise incentives for innovation in pollution control are measured as the amount of the cost savings in the abatement cost if innovation in abatement technology occurs, and it assumes that the introduction of regulations automatically leads to implementation of measures for pollution control.

In practice, in order for enterprises to conduct technical responses to environmental policies, there are several conditions to be met. For large enterprises, technological response to regulations on their own may be possible, but it is difficult for most small- and medium-sized enterprises (SMEs), which are restricted financially and technologically, to satisfy the levels required by environmental regulations. Strict monitoring of each factory is indispensable in order for environmental regulation to be effective. However, it is almost impossible to monitor all SMEs strictly, and introducing environmental regulation alone will not induce all SMEs to take effective measures for pollution control. Therefore, some kinds of financial and technological support for SMEs enterprises are required.

Preferential loans by government-affiliated financial institutions and local government may be mentioned as a typical financial support, and such loans have played an important role in diffusion of pollution control technologies in Japan.[2] As for technical support, local governmental research institutes (LGRIs) have often provided technical guidance to SMEs.[3] As mentioned below, there are various kinds of local governmental research institutes, and it is widely known that the research institutes involved in environmental and sanitation problems have played an important role in monitoring pollutants, comprehending environmental damage and so on. However, the role of industrial research institutes in developing and diffusing pollution control technologies has not been clarified yet.

In this chapter, we analyse, mainly through case studies, the effects of LGRIs on the development and diffusion of pollution control technologies, focusing especially on LGRIs that are involved in manufacturing and that were established for supporting local SMEs by providing technological information and technical guidance. In order to investigate the effectiveness of the technical guidance provided by LGRIs for pollution control, this paper also analyses the cases of technical guidance provided by the regulatory authorities and makes a comparison between them. We hope to clarify a portion of the historical role of LGRIs in the promotion of environmental policies through technical consultation and guidance in Japan and increase knowledge of the conditions under which environmental policies for SMEs are effective.

The construction of this chapter is as follows. In Section 1, we survey the function of external organizations in pollution control and the actual situation of pollution control measures taken by SMEs. In Section 2, we look into the actual situation of LGRIs and survey their role in pollution control. In Section 3, we look into the state of the public research institutes in Tokyo and investigate the effectiveness of the activities of local governmental research institutes. Based on Section 1 and Section 3, in Section 4 we investigate the advantages and problems of local public research institutes and, in Section 5, we summarize and discuss the implications for environmental policies in developing countries.

1.   Role of external organizations in pollution control measures

The difference between SMEs and large enterprises

When SMEs take measures for pollution control in response to environmental regulations they face some obstacles. Among these, the technological difficulties are often pointed out. Financial difficulties have also been noted but they could be subsumed under technological difficulties, considering that the difficulty is in raising money for taking proper technical measures.

The Small- and Medium-sized Enterprise Agency (SMEA) of Japan conducted research on the actual pollution control measures taken by SMEs at the beginning of 1970s. Table 3.1 shows the practical effects of the pollution control measures taken. There are several differences between large enterprises and SMEs. When enterprises face environmental regulations, there are several ways to attain the level mandated by regulation: installing pollution control equipment, changing materials or the process of production and so on. In response

Table 3.1 Effect of pollution control measures

Measures	SMEs	LEs
Installing equipment	47	72
Changing processes	71	72
Changing fuels	79	56

Notes: 1. Figures represent the percentage of respondents who answered 'very effective'.
2. 'LEs' stands for large enterprises.
Source: Compiled by the author from SMEA (1974), p. 234.

to a question concerning the installation of pollution control equipment, whereas 47 per cent of SMEs answered that it was very effective, 72 per cent of large enterprises answered likewise. The reasons why it was less effective for SMEs to install pollution abatement equipment are as follow: insufficient information and knowledge concerning the equipment and technologies, 27 per cent; inadequate equipment capacity due to insufficient finances, 23 per cent; insufficient understanding of the quality and quantity of the pollutants, 19 per cent; and so on (SMEA, 1974, pp. 236–7). Concerning costs, the costs of SMEs tended to be higher than had been expected. This indicates that technical measures for pollution control taken by SMEs did not necessarily work well compared with those taken by large enterprises. What is the cause of these differences between large enterprises and SMEs?

Since the technologies and the systems that are required for pollution control depend on multiple conditions, including the type of pollutants discharged, it is extremely difficult to attain the merits of scale. Thus, pollution control equipment that is appropriate for SMEs tends to be expensive and, moreover, it was not even adequately developed until the beginning of 1970s. The treatment of pollution is complex; that is to say, the effectiveness of the treatment depends on several conditions, so installation of equipment alone is insufficient for effective pollution control measures. In order to control pollution properly, a preliminary investigation of the conditions followed by modification after a period of trial and error are required. Pollution control achieved by changing the process of production is a more fundamental method, and it may in some cases lead to a reduction in the total cost. However, the initial cost of changes in production process tends to be higher, and it is more difficult for SMEs to undertake alone because they are constrained financially and technologically. Therefore, support in many aspects is required from external organizations.

SMEA also conducted research on SMEs' use of external organizations when they took measures for pollution control (SMEA, 1974, p. 236). The result was that 42 per cent of the SMEs used technical guidance provided by the relevant bodies of local governments or LGRIs, and this figure is greater than in the case of large enterprise (29 per cent). On the other hand, when it comes to the pollution control equipment, there is the reverse tendency; that is, while

28 per cent of SMEs used technical guidance from LGRIs, 44 per cent of large enterprises used such guidance. We could say that, at least in the beginning of the 1970s, this result shows that pollution control equipment providers had not necessarily conducted adequate research and development on pollution control technologies for SMEs, and this resulted in an increased role for public research institutes in assisting SMEs in taking proper pollution control measures rapidly in response to environmental regulations.

Role of local governmental research institutes in pollution control

SMEA compiled cases of pollution control by SMEs in the 1970s, and among these, there were some cases in which the technical guidance of LGRIs played an important role in promoting pollution control measures. For example, in nine cases out of the 18 that are treated in SMEA (1975), the role of LGRIs is mentioned. However, the cases in these books were not collected with random sampling; thus, needless to say, this result cannot be directly generalized. Yet, we may say that there was a tendency for LGRIs to play an important role in the promotion of pollution control measures for SMEs.

Among the cases collected, there are those where enterprises asked providers of pollution control equipment to present estimates but the equipment was too expensive to purchase. Thus, with the support of local governmental research institutes, the enterprises attempted to respond to the regulations and finally succeeded in controlling pollution while decreasing the cost dramatically at the same time.

Among these cases, K Ltd in Kagoshima Prefecture presents one of the most typical cases in which LGRIs played a crucial role in promoting measures for pollution control (SMEA, 1975, pp. 64–72). Although K Ltd was required by the regulatory authorities to properly treat wastewater it discharged, it did not have its own technologies for treating wastewater at that time. K Ltd had requested estimates from manufacturers on construction of pollution control equipment, but either the estimates were too expensive or the equipment was of doubtful quality. Thus, K Ltd visited an LGRI for technical guidance, and started to try to treat the problems. In order to design a system for controlling wastewater, it was necessary to know the quantity and quality of the water. Thus, the LGRI repeated water tests and analysis and gathered basic data. Using the data, K Ltd investigated the condition of wastewater discharge and designed a treatment system for wastewater by itself, finally succeeding in controlling its pollution adequately. Although one manufacturer's estimate was about 30 million yen, the actual cost turned out to be 8 million yen. In addition, the actual running cost was approximately a quarter of the estimated running cost.

This case shows how pollution prevention equipment available from manufacturers was sometimes too expensive for SMEs and not necessarily appropriate to their needs, and it was sometimes possible for SMEs to attain the level mandated by regulation through their own efforts and avoid purchasing expensive equipment.[4] Technical guidance from LGRIs functioned to boost the efficiency and effectiveness of the SMEs' efforts. Moreover, the LGRIs' role is not only technology transfer, which already takes place, but also improvement of the ability of SMEs to innovate by themselves.

2.   Actual condition of local governmental research institutes

LGRIs were established in each prefecture and some large cities primarily in order to develop the technological capabilities of regional industries by providing technical guidance, providing tests on request and conducting their own research and development (R&D) activities. Many of the institutes have a long history in their regions. While national research institutes and universities mainly conduct basic research, LGRIs conduct applied research and development, and they have played a large role as organizations for transferring technology to SMEs in each region. While most of the LGRIs are involved in general manufacturing and agricultural research, some prefectures or cities have established special research institutes that reflect the local industrial features. For example, each prefecture in the Tohoku area has a research institute devoted to brewing, and most of the prefectures with a significant textile industry have research institutes devoted to textiles. Starting in the 1960s, each local government has had to tackle environmental problems, and most of the local governments have established a research institute for pollution control or environmental protection. While the research institutes for environmental protection have played a great role in measuring pollution and grasping the effects of environmental damage, it is the traditional research institutes devoted to manufacturing that have usually played the major role in providing technical guidance to SMEs for implementation of pollution control measures.

Tables 3.2, 3.3 and 3.4, respectively, show the number of research institutes, the amount of R&D expenditures, and the number of researchers by sector in Japan. About a half of the research organizations are LGRIs. This is because all prefectures and most large cities have their own research institutes; therefore, LGRIs have at least eased the imbalance in R&D activities among regions in Japan. This contrasts with national research institutes, more than 30 per cent of which are concentrated in Tsukuba area. The number of researchers employed at LGRIs is also high. The R&D expenditure of LGRIs is, however, the smallest among the research institutes.

Since the 1980s, the number of LGRIs has been decreasing. This is a result of the integration of LGRIs since the 1980s, in keeping with the overall trend of administrative and fiscal reform by local government. As Table 3.5 shows, the R&D expenditure per person at LGRIs is far smaller than at other research organizations. This reflects the bad research conditions at LGRIs. Limited R&D

Table 3.2 Institutions performing R&D (natural science only)

FY	Total	Enterprises	Research organizations					Universities
			Subtotal	National	Local	Special corporations	Private	
1965	10,758	9,466	845	83	560	6	196	447
	100	88.0	7.9	0.8	5.2	0.1	1.8	4.2
1970	18,935	17,620	831	88	551	9	183	484
	100	93.1	4.4	0.5	2.9	0.0	1.0	2.6
1975	14,552	13,090	884	84	585	6	209	578
	100	90.0	6.1	0.6	4.6	0.0	1.4	4.0
1980	19,103	17,468	992	89	638	9	256	643
	100	91.4	5.2	0.5	3.3	0.0	1.3	3.4
1985	16,263	14,490	1,078	95	610	8	365	695
	100	89.1	6.6	0.7	3.8	0.0	2.2	4.3
1990	15,792	13,849	1,178	97	584	8	489	765
	100	84.7	7.5	0.7	3.7	0.1	3.1	4.8
1995	16,530	14,485	1,157	97	569	8	483	888
	100	87.6	7.0	0.6	3.4	0.0	2.9	5.4
2000	24,974	22,789	1,016	56	493	9	458	1,169
	100	91.3	4.1	0.2	2.0	0.0	1.8	4.7

Note: The lower value is the percentage of the total.
Source: Ministry of Education, Culture, Sports, Science and Technology, Scientific and Technological Indicators.

Table 3.3 R&D expenditure by sector (natural science only)

FY	Total	Enterprises	Research organizations					Universities
			Subtotal	National	Local	Special corporations	Private	
1965	425,832	252,359	68,426	30,042	24,019	6,722	7,643	105,048
	100	59.3	16.1	7.1	5.6	1.6	1.8	24.7
1970	1,195,328	823,265	154,619	51,560	54,239	34,222	14,598	217,444
	100	68.9	12.9	4.3	4.5	2.9	1.2	18.2
1975	2,621,827	1,684,847	420,699	117,596	111,460	118,959	72,684	516,281
	100	64.3	16.0	4.5	4.3	4.5	2.8	19.7
1980	4,683,768	3,142,256	717,612	185,372	165,966	243,742	122,533	823,900
	100	67.1	15.3	4.0	3.5	5.2	2.6	17.6
1985	8,116,399	5,939,947	1,101,041	227,454	193,052	364,704	316,461	1,075,410
	100	73.2	13.6	2.8	2.4	4.5	3.9	13.2
1990	12,089,593	9,267,166	1,416,079	307,316	252,734	380,637	475,391	1,406,347
	100	76.7	11.7	2.5	2.1	3.1	3.9	11.6
1995	13,191,183	9,395,896	1,920,618	469,044	274,861	605,441	571,273	1,874,668
	100	71.2	14.6	3.5	2.1	4.6	4.3	14.2
2000	14,988,107	10,860,215	2,138,565	483,193	261,385	732,803	661,185	1,989,327
	100	72.5	14.3	3.2	1.7	4.9	4.4	13.3

Notes: The upper value is the amount of expenditure (in million yen).
The lower value is the percentage of the total.
Sources: Ministry of Education, Culture, Sports, Science and Technology, Scientific and Technological Indicators.

Table 3.4 Researchers by sector (natural science only)

FY	Total	Enterprises	Research organizations					Universities
			Subtotal	National	Local	Special corporations	Private	
1965	117,596	58,997	19,466	8,247	9,207	812	1,200	39,133
	100	50.2	16.6	7.0	7.8	0.7	1.0	33.3
1970	172,002	94,060	22,702	8,826	11,149	1,262	1,465	55,240
	100	54.7	13.2	5.1	6.5	0.7	0.9	32.1
1975	255,202	146,604	26,690	9,341	13,732	1,842	1,775	81,908
	100	57.4	10.5	3.7	5.4	0.7	0.7	32.1
1980	302,585	173,244	28,641	9,895	13,988	2,246	2,512	100,700
	100	91.4	5.2	0.5	3.3	0.0	1.3	3.4
1985	381,282	231,097	32,167	10,037	13,994	2,487	5,649	118,018
	100	60.6	8.4	2.6	3.7	0.7	1.5	31.0
1990	484,346	313,948	36,265	10,195	13,713	3,098	9,259	134,133
	100	64.8	7.5	2.1	2.8	0.6	1.9	27.7
1995	574,501	376,639	41,586	10,519	13,724	3,828	13,515	156,276
	100	65.6	7.2	1.8	2.4	0.7	0.7	27.2
2000	643,992	433,758	43,025	10,691	13,822	4,686	13,826	167,209
	100	66.7	6.8	1.7	2.1	0.7	2.1	26.0

Note: The lower value is the percentage of the total.
Sources: Ministry of Education, Culture, Sports, Science and Technology, Scientific and Technological Indicators.

Table 3.5 R&D expenditure per researcher (million yen)

FY	Enterprises	Research organizations		
		National	Local	Private
1970	8.75	5.84	4.86	9.96
1975	11.49	12.59	8.12	40.95
1980	18.14	18.73	11.86	48.78
1985	25.70	22.66	13.80	56.02
1990	29.52	30.14	18.43	51.34
1995	24.95	43.73	20.03	42.27
2000	25.04	45.20	19.10	47.82

Source: Compiled by author from Tables 3.3 and 3.4.

expenditure results in delays in new investment and remodelling of research facilities, which makes it difficult for LGRIs to conduct research and provide technical guidance that reflects the advanced knowledge embodied in the latest facilities. And at the same time, the low R&D expenditure is indicative of a feature of LGRIs, namely that the research activity of LGRIs is catch-up rather than pioneering and, thus, the expenditure required is not necessarily as large as that of other institutes.

3.   Case of the Tokyo Metropolitan Industrial Technology Research Institute

In this section, to provide another example of LGRIs, we treat the case of the Tokyo Metropolitan Industrial Technology Research Institute (TMITRI) and look into the actual condition of its technical guidance and its R&D activities on pollution control by SMEs.

Technical guidance by TMITRI

TMITRI today is the result of multiple integrations over time of public research institutes in Tokyo. Although the Tokyo Metropolitan Government maintains some research institutes other than TMITRI, TMITRI is the largest among them.[5] Table 3.6 shows the outline of TMITRI, the number of researchers, technical guidance, the budgets for R&D and so on. As of 2003, the number of researchers was 158 and R&D expenditure amounted to approximately 1.2 billion yen. However, in the middle of 1970s, the number of researchers exceeded 200.

In Japan, technical consultations and guidance provided to SMEs concerning pollution control increased rapidly in the beginning of the 1970s, when environmental disruptions happened in succession and environmental regulations were rapidly introduced or strengthened. Also in Tokyo, the Metropolitan Government tried to execute its own strict plan for pollution control in addition to the nationwide regulations. In response to the trend, in 1972 TMITRI

Table 3.6   Outline of TMITRI

FY	Number of researchers	Number of consultations	Number of guidances	Number of R&D projects		Budgets for R&D (Thousand)
				Usual	Special	
1970	188	30,653	131	72	7	44,937
1975	202	29,929	254	54	6	146,538
1980	190	31,189	260	52	5	168,295
1985	172	33,945	228	50	4	167,953
1990	167	36,000	250	50	6	172,652

Source: Compiled by the author from TMITRI (1992), pp. 20–3.

established a corner for consultations concerning pollution control technologies.[6] In those days, consultations concerning pollution control accounted for the majority of the consultations. For example, in the field of inorganic chemical industry alone, consultations concerning pollution control amounted to approximately 6,000 per year in the mid-1970s, when the number of total consultations was about 35,000.[7] The number of technical consultations related to pollution control has been about 400–500 recently, and these have accounted for about 1–2 per cent of the total.[8] While most of the consultations have been from SMEs, consultations from large enterprises account for about 10 per cent. Consultations have been requested not only by enterprises subject to environmental regulations but also by manufacturers that provide pollution abatement equipment, although the latter have not been frequent.

TMITRI receives many kinds of technical consultations. The simplest cases are resolved completely by telephone, by telling callers about the technologies capable of treating the existing pollution; however, such cases are not frequent.[9] After consultations, technical guidance is conducted as necessary.[10] In some cases, the equipment installed has not reduced the pollution, although it was supposed, theoretically, to work well. Even if the equipment is the type of end-of-pipe technology that is considered to be comparatively easy to handle, trial and error is often required for it to work well. In other cases, a fundamental transformation of the production process is required. These treatments are more than a simple transfer of pre-existing technologies. In addition to conducting technical guidance responding to individual needs, TMITRI has also provided several training courses in order to improve the technical ability of SMEs in general.

One factor that enables LGRIs to conduct proper technical guidance is that they conduct their own R&D and improve their knowledge. In TMITRI, ordinary R&D and special R&D are conducted. While the former is determined by considering the administrative needs for research in Tokyo and the requests of industry groups, the latter type of R&D is for the purpose of resolving the

Table 3.7 Number of ordinary R&D projects

FY	Energy saving	Pollution control	Quality of products	New products	Total
1971	10	6	27	33	76
1972	2	8	29	13	52
1973	8	12	20	14	54
1974	4	13	22	14	53
1975	7	12	21	14	54
1976	10	12	16	16	54
1977	7	8	17	18	50
1978	8	6	20	16	50
1979	8	4	23	17	52
1980	13	3	19	17	52
1981	10	3	21	19	53
1982	10	2	21	18	51
1983	13	2	18	17	50
1984	6	1	19	24	50
1985	9	2	18	21	50
1986	8	2	16	24	50
1987	6	1	19	24	50
1988	4	2	21	23	50
1989	3	2	20	25	50

Source: TMITRI (1992), p. 72.

problems that SMEs commonly have not only in Tokyo but all over Japan, and the cost is expected to be relatively high. Some of the projects have been provided with subsidies from the national budget. For urgent problems, the governor has occasionally issued directives concerning the subject of the research.[11] Table 3.7 shows the trends in ordinary R&D. Until the end of the 1970s, development of pollution control technologies had been regarded as the most urgent R&D topic. R&D for pollution control was conducted frequently then but, since the 1980s, there have been only one or two R&D projects on pollution control. After the first oil shock in the mid-1970s, R&D for energy saving became one of the major fields. A similar trend was seen in enterprises.

Cases of technical guidance for the plating industry

The plating industry, which is a typical pollution-intensive industry and is composed mostly of SMEs, is concentrated in the Tokyo area and it is, so to speak, the local industry of Tokyo. TMITRI has often provided technical guidance to the plating industry since around 1950 (The Association of Tokyo Plating Industry, 1990, pp. 324–5). This technical guidance begun by TMITRI had great impact on nationwide technology policies for SMEs. For example, the Small-and Medium-sized Enterprise Agency introduced subsidies for prefectures in order to promote technical guidance by each prefecture. After concerns over environmental problems were raised, technical guidance for the plating industry came to focus mainly on the treatment of wastewater. Through such detailed efforts, TMITRI has maintained a close relationship with the plating industry groups. Since it has become apparent that the trend is towards stronger environmental regulations, TMITRI has often received the consultations from the plating industry groups that want to attain the target set by environmental regulations.[12] The Association of Tokyo Plating Industry itself acknowledges that the technical guidance of TMITRI has been very effective.[13]

The plating industry has had a close relationship with not only TMITRI but also the relevant authorities. In the beginning of the 1970s, many of plating factories were interspersed among houses in the city, and it became difficult to continue operation because of the load they placed on the environment. However, many of plating factories were small and technologically and financially restricted, and thus it was difficult for each factory alone to decrease the load. Jointly responding to administrative guidance from the Tokyo Metropolitan Government, many of plating factories decided to move to a landfill area in order to treat their wastewater collectively. At the same time, the plating industry group also decided to construct a collective facility to treat the cyanide in the wastewater discharged from each factory. In order to implement these projects properly, parties concerned formed a special committee, mainly composed of the relevant authorities and the industry group. TMITRI was also involved in the committee as a technical adviser. In those cases, several types of assistance were provided by Tokyo Metropolitan Government.[14]

There are cases where innovative technologies were developed through the close relationships between TMITRI and the industry groups. Among recent environmental regulations for the plating industry, the emission standard for boron was strengthened to an extremely strict level, making it impossible to attain the goal with the traditional technologies. TMITRI received a request from the plating industry and started engaging in R&D. After trial and error, the institute succeeded in developing technologies for abating the boron in wastewater as well as a new method that does not use boron in the production process. The former is an example of end-of-pipe technologies and the latter is an example of cleaner production technologies. Both were patented and it is particularly noteworthy that the latter is the first method in the world that does not use boron in the production process.

Role of the regulatory authorities in technical guidance

Although we have discussed the role of LGRIs, not only LGRIs but also the regulatory authorities have provided technical guidance for implementation of pollution control measures in practice. From the beginning to the middle of the 1970s, when urgent responses to strengthened environmental regulations were required, the engineers of the regulatory authorities were also playing an important role in assisting SMEs to take pollution control measures by providing them with technical guidance.[15] In this period, it seems there was a tendency

Table 3.8   Number of engineers at Pollution Control Bureau

FY	Secretariat	Engineers			Total
		ERI	Except ERI	Total	
1970	162	51	131	182	505
1972	296	54	187	241	550
1974	377	62	274	336	735
1976	362	58	273	331	712
1978	343	54	268	322	689
1980	330	53	260	313	661
1982	321	51	259	310	642
1984	303	49	254	303	611

Note: ERI stands for Environment Research Institute.
Source: Compiled by the author from the annual report of the
Pollution Control Bureau, Tokyo Metropolitan Government.

for the regulatory authorities to also play an important role in technical guidance for pollution control not only in Tokyo, but also in other regions.[16] Table 3.8 shows the fluctuations in the number of engineers in the Pollution Control Bureau of the Tokyo Metropolitan Government. In the mid-1970s, the number of engineers assigned to the bureau, excluding the engineers and researchers of the Environment Research Institute, reached 280, which accounted for 38.5 per cent of the employees and was larger than that of TMITRI. This technostructure enabled the authorities to conduct effective technical guidance. In addition to the original technical guidance that LGRIs provided on their own, when pollution control authorities provided administrative guidance, LGRI researchers sometimes examined the technical possibilities of the guidance upon request from the regulatory authorities.[17] However, there were few cases where the regulatory authorities formally asked TMITRI to examine the technical possibility of their administrative guidance, and the collaboration between them was limited to the level of relationships between individual researchers and engineers in many cases. This means that LGRIs and the regulatory authorities have usually conducted technical guidance independently.

Once the urgent responses of enterprises to environmental regulations had been implemented, the role of the regulatory authorities in providing technical guidance to SMEs gradually decreased. There were the cases in which the engineers who had been hired by the regulatory authorities were reshuffled to LGRIs.[18] It is true that the number of engineers at the bureau decreased after the mid-1970s, but their proportion to the total number of employees did not decrease, as shown by Table 3.8. Therefore, the ability of the control bureau to provide technological guidance did not decrease markedly. Once urgent responses to environmental regulations had been implemented, it seems that there was a tendency for the regulatory authorities' role in providing technical guidance to diminish in areas other than Tokyo as well. This problem will be discussed below.

4.   Advantages and problems of local governmental research institutes

Advantages of local governmental research institutes

As mentioned above, the role of LGRIs varies from offering technology information to providing technical guidance based on advanced R&D activities conducted on their own. Since information-gathering cost was extremely high for SMEs in the 1970s, the SMEs particularly welcomed even simple functions such as provision of recent technology information by LGRIs. However, the diffusion of the Internet has sharply decreased the information-gathering cost. It is true that the need to provide technology information will not vanish as long as there is a technical disparity between LGRIs and SMEs, and this disparity will not easily disappear,[19] but the importance of the simple function of provision of technology information by LGRIs is unavoidably diminishing. Considering this situation, it seems that what is required of LGRIs most today is technical guidance based on their advanced R&D activities and accumulated experience.

With respect to technical guidance, what advantages can LGRIs offer to SMEs compared to universities and national research institutes? Because one of the main purposes for which LGRIs were established is to contribute to the technical progress of SMEs in each region, it is comparatively easy for SMEs to use LGRIs' services. In comparison, many SMEs feel that the threshold is high at universities and national research institutes. Thus, as described above, many SMEs visit LGRIs more often than universities and national research institutes for technical guidance. In order to respond to the consultation properly, data related to trade secrets, such as the detailed contents of material discharged into water, are sometimes required. LGRIs conduct tests on request as one of their services; therefore, it is comparatively easy for LGRIs to obtain such data.[20] Thus, instead of the regulatory authorities, LGRIs are more appropriate organizations for providing technical guidance.

The career paths of researchers in LGRIs have also facilitated the formation of close relationships with SMEs. Since researchers at LGRIs in Japan typically spend their entire careers with one institute, it is relatively easy for researchers at LGRIs and firms in the region to develop a close relationship.[21] The accumulated experience of LGRIs in providing technical guidance strengthens their ability as organizations for technical guidance. Although the number of consultations on pollution control has been decreasing recently, in the process of responding to consultations from SMEs, LGRIs construct, so to speak, a practical 'database' for pollution control measures. The word 'database' is also used as a metaphor here; that is, it includes the experience accumulated by researchers at LGRIs, which is not necessarily compiled in explicit form as documents or electronic media. The existence of this 'database' has improved the ability of LGRIs to treat problems. That is, there is a positive feedback loop. Although a certain level of research ability is a prerequisite, that alone would not necessarily be sufficient to solve SMEs' pollution control problems. It could be said that the existence of the 'database' accumulated through the long experience is the maximum advantage of LGRIs as organizations for technical guidance, although other research institutes such as national research institutes and universities in general possess a higher level of research ability.[22]

Differences between the technical guidance of LGRIs and the authorities

As mentioned in 'Role of the regulatory authorities in technical guidance', in the period when environmental regulation was rapidly strengthened, not only LGRIs but also the regulatory authorities played an important role in technical guidance for pollution control. Were there any differences between the roles of the two in technical guidance?

Since LGRIs are, of course, research institutes, their ability to perform research is, in general, superior to that of the regulatory authorities. Thus, when advanced treatments were required, most firms asked for the advice of LGRIs. In the cases where the required level of technical guidance is not so advanced, the regulatory authorities are also capable of responding to the consultations. As mentioned above, data related to trade secrets are sometimes required in order to conduct appropriate technical guidance for pollution control. In certain cases, there were some firms that hesitated to offer the regulatory authorities important data related to trade secrets for fear that the regulations would be strengthened using the data offered. These firms visit LGRIs rather than the regulatory authorities for technical guidance, even in the cases where the regulatory authorities could sufficiently treat the problems. However, if the regulatory authorities, which have the power to change the strictness of regulations such as those on emission levels, were trusted not to introduce regulations that would impose a heavy burden on SMEs, they could play a role similar to that of LGRIs, judging from the cases in Tokyo and Nishi Yodogawa Ward.[23] Consequently, if the technostructure of the regulatory authorities is high enough to conduct technical guidance (even though their research ability is not as high as that of LGRIs) and if they could obtain the trust of the SMEs in the region, then which organization the SMEs would choose to depend upon when they require technical guidance would be mostly determined by their relationship to date with the respective organizations and the level or difficulty of the technical guidance required. We could say that there are basically few differences between the roles of the two at the initial stage of taking measures. This indicates that the most advanced knowledge is not necessarily required in order to provide appropriate technical guidance in many cases, and it is important that there exists an organization in each region where both information about pollution control and technical consultations are readily available, even if the organization is not in the form of a research institute.

After the first round of introducing or strengthening of environmental regulations was completed, there were not so many options for companies to choose from when it came to technical treatments that were comparatively easy to introduce; that is, there were not many treatments available with low marginal cost. Therefore, it was natural that the regulatory authorities' role in technical guidance for pollution control measures decreased after the mid-1970s.

Recent problems at local governmental research institutes

There have been some recent problems that may undermine the ability of LGRIs to provide technical guidance. The first problem that should be pointed out is that the number of LGRIs has been decreasing as a result of the integration of LGRIs. As mentioned above, it is important that institutes that are capable of conducting technical guidance exist in each region; thus, the decrease in LGRIs themselves might be calamitous. If the number of LGRIs decreases due to the integration, one option is to open branch offices where the antecedents of integrated LGRIs used to be. In that case, even if some of LGRIs are integrated, the places where technology information is easily available would not necessarily decrease at once. In the long term, however, some of the branches might be closed.

In a similar vein, another problem that can be pointed out is that the number of experienced researchers at LGRIs has been decreasing, following the overall tendency in Japan towards staff reductions due to local government budget constraints. Although the number of researchers is decreasing, LGRIs are required to play new roles, such as coordinating collaborations between universities and industry, in addition to their traditional roles. Collaboration between universities and industry itself is important and, in order to involve SMEs in the collaboration, LGRIs are appropriate coordinators because they generally grasp the technical and economic conditions of SMEs in the region. However, the purposes of the collaborations now in progress are not really suited to most SMEs but are better suited to advanced enterprises that are competitive. In practice, while there are a few competitive SMEs, many SMEs find it difficult to attain the level required by environmental regulations without the support of LGRIs. Adding new roles to LGRIs could make it difficult for them to execute their traditional roles, like technical guidance, thereby decreasing the effectiveness of environmental policies for SMEs.

LGRIs themselves sometimes admit that, when it comes to providing technical guidance that requires experience, the level of the technical guidance provided has decreased (TMITRI, 1992, p. 62.). As mentioned above, if the technical guidance of LGRIs is effective, the effectiveness can be attributed to the contribution of the large amount of know-how accumulated in the process of conducting various types of technical guidance. Experience needs to be accumulated as an organization in order to conduct effective technical guidance. Complete documentation of know-how is impossible because some of it can be transmitted only through face-to-face communication. If new researchers are not assigned to LGRIs and the average age of the staff rises, it may be difficult to transmit to the next generation the know-how that is said to be the most valuable asset of LGRIs. It is often said that many SMEs in Japan do not have successors to carry on the business, and the transmission of manufacturing skills to the next generation is difficult. In many LGRIs, if new hiring of researchers is suppressed, a similar problem may occur.

5.   Concluding remarks: implications for developing countries

In this chapter, we have discussed the role of local governmental research institutes in assisting SMEs with their pollution control measures, a matter to which sufficient attention has not been paid so far. It was confirmed that LGRIs are able to assist SMEs with the implementation of measures for pollution control, and this assistance can improve the effectiveness of environmental policies in Japan. Finally, let us consider the implications of Japan's experience with LGRIs for developing countries.

One of the most fundamental reasons why LGRIs could be established in many regions in Japan was that Japanese society was capable of supplying sufficient researchers and engineers to local research institutes. This capability is highly dependent on the national education and innovation system, and it is almost impossible to establish such a system in the short term. One of the reasons why environmental policies are not always effective in developing countries is that there is an insufficient number of researchers and engineers who are engaged at the field level in implementing pollution control measures. Consequently, the first requirement in developing countries is to construct and improve the education systems, which supply the researchers and engineers. This is a prerequisite for establishing research institutes like the LGRIs in Japan in other developing countries.

The next question is, if sufficient researchers and engineers are supplied, then is it possible to establish institutes that can conduct effective technical guidance for pollution control in each region? It is possible that local public research institutes may not be able to secure a sufficient number of engineers even if the supply of engineers is sufficient on a nationwide scale. However, even in the case that local research institutes cannot offer favourable terms, they could secure excellent engineers if the social credibility of local government or local research institutes is high enough. As mentioned above, the social credibility of LGRIs is also crucial for the effectiveness of technical guidance. Even if it is possible to establish an organization like the local public research institutes in Japan and arrange for researchers and engineers, the institute will not be able to obtain sufficient data and conduct effective technical guidance for pollution control if the institute is not socially trusted.

Thus, it is necessary not only to establish and maintain the facilities, but also to ensure that the social situation is such that SMEs can easily use the services of local institutes for pollution control.

Notes

1. Examples of this type analysis are Zerbe (1970), Downing and White (1986) and Milliman and Prince (1989). Among them, it could be said Milliman and Prince (1989) is the most detailed and comprehensive analysis. Most of this type of theoretical analysis indicates that economic instruments like the environmental taxes exert a stronger incentive to engage in research and development than direct regulations.
2. For example, the OECD (1975) evaluates the effectiveness of the preferential loan in Japan. Japan Environmental Corporation (JEC), which was typical of organizations that provided preferential loans, not only provided preferential loans, but also conducted technological studies concerning the equipment which was the object of the preferential loans. See Akira Morishima, *Kankyo Jigyodan no Keii* (*History of Japanese Environmental Corporations*) (interview, in Japanese), in Research Group on Environmental Policies in Asia (2002), pp.119–36.
3. In economics, whether public knowledge accumulated in public research institutes (e.g., LGRIs, national research institutes and universities) directly or indirectly promotes industrial innovation has provoked a controversy from both theoretical and empirical viewpoints. Jaffe (1986) studied the effects of distance between industry and public research institutes on innovation by industry using patent data and asserted that the proximity of a company to public research institutes bears closely on the benefit from spillovers of R&D activities. This result shows that, while spillovers from public research institutes promote innovation of firms, the spillover is geographically limited. Similar studies are conducted and developed in, for example, Acs *et al.* (1992), Autant-Bernard (2001).
4. Needless to say, this does not mean it is always possible.
5. Although the name of the nucleus of the TMITRI's antecedent was not the same before the integration, we also call the institute before the integration TMITRI in this chapter in order to avoid confusion.
6. Not only TMITRI but also most of the LGRIs in other regions implemented special measures for promoting pollution control technologies for SMEs in those days.
7. TMITRI (1992), p. 60. This figure is the number of consultations concerning actual pollution control. Consultations concerning the development of environmentally friendly products are not included in this figure. Thus, consultations concerning environment were more frequent.
8. In 2003, the number of the technical consultations for pollution control received by TMITRI was 473, which is just 1 per cent of the total. Annual Report of TMITRI (2003), p. 61.
9. Personal communication with researchers at TMITRI (19 January 2004).
10. After 1981, the technical adviser system was introduced. When the technical guidance that is required is expected to take a long time, the enterprise that needs it contracts the adviser, who registers with TMITRI, and TMITRI shares half the cost.
11. For example, concerning the PCB problem, Ryokichi Minobe, the governor of Tokyo, directed each research institute of the Tokyo Metropolitan Government to engage in research to find a technical solution. See TMITRI (1992), p. 61.

12. Concerning the plating industry in Tokyo, technical consultations in response to environmental regulations are often brought in through the industry group.
13. The Association of Tokyo Plating Industry (1990), pp. 324–6. In this reference, technical guidance by the antecedent of TMITRI is often mentioned. Concerning technological development, in general, industry groups tend to undervalue the role of public institutes and to emphasize their own efforts; therefore, the role of the antecedent of TMITRI is actually thought to be crucial.
14. A plating enterprise called A. Ltd, which participated in the collective relocation to landfill area, said that the relocation and the change of the production process there could not have succeeded without repeated assistance by Tokyo Metropolitan Government. Personal communication (11 May 2004).
15. Personal communication with researchers at TMITRI, as mentioned above, and Mr Kazuo Hishida, who was formerly an official of the Tokyo Metropolitan Government and who had long been engaged in fieldwork supervision and technical guidance for pollution control measures.
16. Nishi Yodogawa Ward, which is located in the northwestern part of the City of Osaka, is mentioned as another example of this situation. Since Nishi Yodogawa Ward has housing that intermingles with small and medium-sized factories, several cases of environmental damage happened. The City of Osaka organized the Special Task Force for Pollution in Nishi Yodogawa Ward in order to ease pollution urgently in 1970. The Task Force was composed of 13 staff members who were all engineers and specialized in detailed research on pollution sources and who offered technical guidance for pollution control. A separate analysis group of three staff members was organized to support the Task Force by examining the relationship between the concentration of air pollution in the area and the data on the amount of air pollutant emissions from each factory obtained through spot inspections by the Task Force. The group sent the findings to the Task Force to serve as a guideline for its activity. The Task Force sometimes provided 'top down' technical guidance, but the members of the Task Force also sought technical measures together with engineers from factories to achieve the levels required by environmental regulations. The City of Osaka introduced preferential loans for SMEs to install pollution abatement equipment. When SMEs in Nishi Yodogawa Ward applied for the loans, the Task Force examined the effectiveness and provided technical guidance. Due to its attitude, the Task Force could obtain the trust of many factories in Nishi Yodogawa Ward, and there were only a few cases of trouble between the Task Force and the factories in this area. The Special Task Force achieved its goal and was disbanded in 1973. The members of the Task Force, except the analysis group, were hired by the regulatory authorities, not as researchers at LGRIs. See Kuroda (1996), pp. 211–81 and World Bank (1996), pp. 43–53.
17. Personal communication with the researchers at TMITRI, as mentioned above. Public research institutes other than TMITRI were also involved in technical guidance.
18. Personal communication with researchers of TMITRI, as mentioned above. There was also a reshuffle among the research institutes.
19. TMITRI set up a corner in 1969 with equipment catalogues from many enterprises, regardless of pollution control technologies. At the end of the 1990s, this corner was closed but there were many requests from SMEs for the corner to be continued. This episode shows that it is still expensive for SMEs to gather technology information, even today, and, thus, the function of LGRIs in providing technology information is still important.
20. Personal communication with researchers at TMITRI, as mentioned above.

21. On the other hand, there are shortcomings in a career system where few reshuffles of researchers occur. The technological and research skills of the staff might become obsolete and the low rate of turnover of the staff makes it hard to bring in new researchers with high skills in emerging technology and research areas. See Shapira (1996), pp. 317–34.
22. Needless to say, both universities and national research institutes have provided technical guidance to SMEs and some of them have been effective. However, technical guidance for SMEs is not the main purpose of the institutes; thus, how frequently researchers provide technical guidance depends upon the characteristics of the individual researchers.
23. According to Mr Hishida, mentioned above, he always tried not only to abate pollution but also improve the productivity as much as possible in order to decrease the burden even a little when he conducted technical guidance for pollution control.

References

Acs, Z. J., D. B. Audretsch and M. P. Feldman (1992) 'Real Effects of Academic Research: Comment', *American Economic Review*, 82, 363–7.

Association of Tokyo Plating Industry (1990) *Hyaku-syu-nen kinen-shi (A Hundred Years' History of the Association of Tokyo Plating Industry)* Tokyo (in Japanese).

Autant-Bernard, C. (2001) 'Science and Knowledge Flows: Evidence from the French Case', *Research Policy* 30(7), 1069–78.

Cockburn, I. and R. Henderson (1997) 'Public–Private Interaction and the Productivity of Pharmaceutical Research', National Bureau of Economic Research Working Paper, No. 6018.

Downing, P. B. and L. J. White (1986) 'Innovation in Pollution Control', *Journal of Environmental Economics and Management*, 13, 18–29.

Fujikura, Ryo (2002) 'Nihon no chiho-kokyo-dantai no iou-sankabutsu taisaku: kodo-kjeizai-seichoki ni jissisareta kogai-boshi-kyotei to gyosei-shido,' (Local Measures in Japan for Controlling Sulphur Oxide Emissions: Pollution Control Agreements and Administrative Guidance Adopted during the Era of Rapid Economic Development), in T. Terao and K. Otsuka (eds) *Dynamism of the Policy Process in 'Development and Environment': The Experiences of Japan and Problems in East Asia*, Tokyo, Institute of Developing Economies, 37–78 (in Japanese).

Hashimoto, Michio (1987) *Shi-shi kankyo gyosei-shi (Private History of Environmental Policies)* Tokyo, Asahi Shimbun-Sha (in Japanese).

Jaffe, A. (1986) 'Technological Opportunity and Spillovers of R&D: Evidence from firms' patent, profits, and market value', *American Economic Review*, 76(5), 984–1001.

Kuroda, Takayuki (1996) *Toshi sangyo-kogai no genten-nishi yodogawa kogai (Nishi Yodogawa: Starting Point of Industrial Pollution)* Tokyo, Doyukan (in Japanese).

Milliman, S. R. and R. Prince (1989) 'Firm Incentives to Promote Technological Change in Pollution Control', *Journal of Environmental Economics and Management*, 17, 247–65.

OECD (1975) *Environmental Policies in Japan*, Paris, OECD.

Osaka City Government (1972) *Nishi yodogawa taiki-osen kinkyu taisaku jisshi hokoku (Report on Urgent Measures for Controlling Air Pollution in Nishi Yodogawa Ward)* Osaka (in Japanese).

Research Group on Environmental Policies in Asia (2002) *Nihon no kogai keiken ni kansuru hiaringu no kiroku (Record of Interviews Concerning Experiences of Pollution Control in Japan)* Chiba, Institute of Developing Economies (in Japanese).

Shapira, P. (1996) 'Modernizing Small Manufacturers in the United States and Japan: Public Technological Infrastructures and Strategies', in M. Teubal *et al.*, *Technological Infrastructure Policy*, Dordrecht: Kluwer Academic Press.

Small- and Medium-sized Enterprise Agency (SMEA) (1974) *Chu-sho kigyo hakusyo (White paper on small- and medium-sized enterprise)* Tokyo (in Japanese).

Small- and Medium-sized Enterprise Agency (SMEA) (1975) *Chu-sho kigyo kogai boshi kaizen jireishu (Case Collection of Improvement of industrial Pollution of SMEs)*, Tokyo (in Japanese).

Sukegawa, Nobuhiko (1991) *Kankyo mondai to jichitai (Environmental Problems and Municipalities)*, Tokyo, ZOION-sha (in Japanese).

Suzuki, Shigeru (2000) *High-tech-gata kaihatsu seisaku no kenkyu (Studies on Development Policy of High-tech Industry)* Kyoto, Minerva (in Japanese).

Terao, Tadayoshi, and Kenji Otsuka (eds) (2002) *'Kaihatsu to kankyo' no seisaku-katei to dainamizumu: nihon no keiken, higashi-ajia no kadai (Dynamism of Policy Process in 'Development and the Environment': The Experiences of Japan and Problems in East Asia).* Chiba, Institute of Developing Economies (in Japanese).

Teubal, M., D. Foray, M. Justman and E. Zuscovitch (1996) *Technological Infrastructure Policy,* Dordrecht, Kluwer Academic Press.

Tokyo Metropolitan Industrial Technology Research Institute (1971–2002) *Annual report of Tokyo Metropolitan Industrial Technology Research Institute* Tokyo (in Japanese).

Tokyo Metropolitan Industrial Technology Research Institute (1992) *Tokyo-toritsu kogyo-gijutsu center niju-nenshi (Twenty Years' History of Tokyo Metropolitan Industrial Technology Research Institute)* Tokyo (in Japanese).

World Bank (1996) *Japan's Experience in Urban Environmental Management. Osaka: A Case Study.* Washington DC.

Zerbe, R. O. (1970) 'Theoretical Efficiency in Pollution Control', *Western Economic Journal*, 8, 364–76.

4
Administrative Guidance of Japanese Local Government for Air Pollution Control

Ryo Fujikura

Introduction

Administrative guidance (*Gyoseishido*) has been conducted in the various fields in Japan. During the post-war period, the Ministry of International Trade and Industry (MITI, currently the Ministry of Economy, Trade and Industry) frequently issued administrative guidance in order to promote infant industries, protect declining industries, rearrange the national industrial structure and control public enterprises (Oyama, 1996, p. 44). While MITI only partially succeeded in attaining these objectives, the guidance has often been lauded as a successful soft tool adopted by the national government that contributed to the remarkable development of the Japanese economy. On the other hand, the accusation has often been made that the guidance is a symbol of the opacity and ambiguity of the Japanese administration, and this assertion has been made particularly by the United States government, which is impatient with the trade imbalance between the two countries (Shindo, 1992).

Administrative guidance has also been issued by Japan's local governments to local industries, with one of the objectives being to control industrial pollution. In December 1970, the national government finally adopted a substantial pollution control policy because most of the industrial cities faced serious pollution. However, it was not until the late 1970s that the results of the new policy began to appear and the industrial pollution problems were substantially solved. In some industrial cities, such as Osaka and Kitakyushu, however, improvement of the situation started during the early 1970s. Administrative guidance given by the municipal governments in those cities played a central role in pollution control.

The Osaka Municipal Government of Osaka Prefecture issued administrative guidance that was based neither on the national laws nor on local ordinances. No sanction against non-compliance with this guidance was stipulated. In Kitakyushu City of Fukuoka Prefecture, the guidance was provided based on pollution control agreements (PCAs) concluded between the municipal government and local industries. No sanctions were stipulated in the PCAs, either. The guidance in both cities was mainly provided orally, in a rather unofficial manner. No obvious compelling power to enforce the compliance seems to have existed. Despite its unclear legitimacy, administrative guidance in both the cities successfully reduced pollution from stationary sources, particularly sulphur oxide emissions, during the early 1970s.

A number of studies were conducted on administrative guidance issued by the Japanese government concerning national industrial and financial policies. There are also several studies on the Japanese experience with industrial pollution during the 1960s and 1970s. However, very few studies have been conducted on the administrative guidance issued by the local governments for industrial pollution control. This chapter examines the guidance issued by the Osaka and Kitakyushu municipal governments, with focus on sulphur oxide emissions control. First, administrative guidance in Osaka and Kitakyushu will be described in the following two sections. Then, factors that enabled the municipal governments to control pollution will be identified. Finally, an evaluation will be made of the guidance from the viewpoint of environmental effectiveness, efficiency, transparency and implications for technical cooperation.

1.   Osaka City

Social background

Despite the lack of any legal basis for it or sanctions against non-compliance with it, local industries adopted costly pollution control measures in accordance with the administrative guidance issued by the Osaka Municipal Government. The specific social background of Osaka City was what made the guidance effective and compelled the local industries to follow it.

Since the seventeenth century, Osaka City had held a special status among municipalities and had been granted various types of autonomous authority by the Japanese feudal government. Since that time, the local communities and industries have played an important role in the 'management' of the Osaka municipality. Since the post-war period, the role of municipal management has been undertaken by local development associations (LDAs) established in each administrative ward. The LDAs have been cooperating with their municipal ward office to enforce Osaka City's local policies. The LDA has a pyramidal structure: the LDA consists of local community unions, which in turn consist of local communities. The local communities are further made up of local citizens' 'groups', and every family belongs to one local group. Merchants and factory owners are influential in local politics, and they are often appointed as directors of these local organizations.

Due to this traditional cooperation between the municipal government and the local communities, the citizens view the municipal government as their own organization for managing their communities. The directors of the local organizations feel that they themselves 'administer' Osaka City. As a result, the citizens are seldom in conflict with the municipal government.

This cooperative relationship results in continuous donations by local businesses and citizens to Osaka City, which is unusual in other Japanese municipalities. In turn, the municipal government provides support to local industries, particularly small and medium industries (SMIs), in various ways. For example, the municipal government has been financially supporting local SMIs through soft loans. Also, the Osaka Municipal Institute of Technologies has granted patents to SMIs without charge. This mutual relationship between the local businesses and the municipal government facilitated the implementation of effective measures for industrial pollution control in Osaka City.

From the mid-1950s, soot and smoke started to annoy citizens as the economy recovered from the destruction of the Second World War. Most of the air pollution problems during that period were caused by coal dust. In the wintertime, thick smog emitted from building heating systems covered the downtown areas. The smog sometimes became so thick that it blocked sunlight, and car drivers had to turn on their headlights even in the daytime. Consequently, it became clear among local businesses that it was important to adopt air pollution control measures in order to protect human health and to improve the city's environmental amenities.

The municipal government encouraged the local businesses to organize Soot and Smoke Prevention Cooperation Associations, and the first association was established in 1958 in Higashi Ward. Afterward, an association was established in every ward, and the Osaka Alliance of these associations was established in 1960. These organizations conducted investigations of anti-pollution technologies and raised the awareness of the business community. Through this mechanism, the municipal government was able to guide local factories, although there was no legal basis for such guidance.

The municipal government was dismayed when the Soot and Smoke Control Law was established in 1962. This law delegated the authority for air pollution control to the prefectural governors, and the municipal mayor was not delegated any authority to control pollution.[1] This eventually clarified the fact that the guidance of the municipal government to factories lacked legal basis; nonetheless, the mayor continued to issue guidance.

Almost all the local factories followed the guidance of the municipal government and allowed inspections by the local officials without questioning the legal basis of such activities. There were only a few factories that refused the inspections. Even the power plants and gas plants permitted the inspections and followed the guidance, although they were not subject to the Soot and Smoke Control Law. (They were controlled directly by MITI based on other laws.) The municipal government was the largest stockholder of the Kansai Electric Corporation. Osaka Gas Corporation was originally the Municipal Gas Department and was privatized due to the establishment of the Gas Corporation Law. With this background, the municipal government was able to control air pollution from the power plant and the gas plant without any difficulties.

When the Soot and Smoke Control Law was fully enacted in 1963, the Osaka Prefectural Government delegated the authority for the enforcement of the law, including inspection, to the municipal government. The law allowed the prefectural governors to delegate their authority to the municipalities, but prefectural governments usually resisted transfer of their authority to municipal governments. In this case, nonetheless, the prefectural government recognized the effectiveness of implementing anti-pollution measures through the municipal government and was, therefore, convinced to delegate its authority (Fujikura, 2005).

Administrative guidance

Until 1969, when a national environmental quality standard (EQS) for sulphur oxides (sulphur dioxide and sulphur trioxide) was adopted by the Cabinet,[2] there was no national target for pollution control. The Ministry of Health and Welfare drafted the Bill for the Standards for Prevention of Living Environment Pollution in 1955. It stipulated the 'allowable limit' of sulphur dioxide at 0.1 ppm. However, the ministry failed to submit the bill to the Diet due to strong opposition from the business community. The Soot and Smoke Control Law established in 1962 was able to prescribe emission standards but not an EQS.

In the absence of a national EQS, the municipal government established its own environmental target as early as the 1960s. The city organized the Osaka Municipal Council for Pollution Control and obtained the council's report, entitled 'Environmental Management Standards regarding Air Pollution', in 1965. The standards included ambient air quality standards for sulphur oxides of 0.1 ppm for a daily average and 0.2 ppm for a maximum one-hour average. These standards were set with consideration for the criteria established by the World Health Organization (WHO) and their feasibility. The municipal government established an administrative guidance plan that incorporated the standards (Osaka Municipal Government, 1994).

The municipal government held a number of meetings to thoroughly convey to the local business community its pollution control policy, as well as to convey information about local air pollution, health damage occurring in the city, administrative guidance, anti-pollution technologies, availability of soft loans, a factory relocation programme as an anti-pollution measure, tax exemptions, supply of low-sulphur heavy oil and so on. The meetings were held at every opportunity, including at the conferences of the Chamber of Commerce and industry associations, factories' meetings, commercial facility meetings and meetings of the Municipal Soot and Smoke Prevention Cooperation Association.

Factory personnel, including those from SMIs, who attended the meetings understood the necessity of adopting air pollution control measures and were able to prepare their own anti-pollution plans. The municipal government presented a concrete target for emission reduction for individual facilities based on solid scientific data and the local environmental management standards. In order to ensure that the factories attained their targets, the municipal government requested that the factory personnel each submit an individual plan for air pollution prevention. Most of the SMIs did not possess adequate knowledge, and the municipal government provided technical advice. Draft plans submitted to the municipal government were examined by the municipal officials. No plan was accepted until the planned measures were regarded as satisfactory by the municipal government, and plans were often returned to the factories for revision.

Even when a final individual plan was submitted, the municipal government did not issue any official document to acknowledge the plan. The municipal government regarded the submission of a plan as a voluntary action. Once an individual plan was accepted by the municipal government, every factory actually adopted the measures stipulated in the plan. Through this individual guidance, the municipal government succeeded in cutting down air pollution.

The Blue Sky Programmes

In 1972, MITI conducted a comprehensive study on industrial pollution (CSIP) in the Osaka area, including a wind tunnel experiment, and forecast air pollution in Osaka Prefecture based on emissions data of 95 large factories in the prefecture. It was predicted that the maximum concentration of sulphur oxides on the ground could be reduced by approximately 30 per cent to 40 per cent if the heavy oil consumed in these factories was replaced by an oil with a sulphur content of less than 1.7 per cent.[3] Then, the prefectural government established the First Blue Sky Programme in June 1969, and the municipal government issued administrative guidance for large factories to use the low-sulphur heavy oil.

However, in downtown Osaka, building heating systems significantly contributed to local air pollution. It seemed that attaining the EQS for sulphur oxides would be impossible even if all large factories attained their emission reduction targets and the other facilities complied with the emission standards of the Air Pollution Control Law established in 1968.[4] Thus, the municipal government established the Second Blue Sky Programme in October 1969. This programme was aimed at reducing emissions from non-production facilities, which were not subject to the national air pollution control law. The municipal government instructed owners of buildings located in the downtown areas to switch their fuel to heavy oil containing less than 1.0 per cent sulphur. Except for a few cases, most of the building owners were cooperative and switched fuel once they understood the necessity of the measure.

Emergency measures against air pollution in Nishi Yodogawa Ward

In Nishi Yodogawa Ward of Osaka City, small factories were built close to residential houses and there were numerous complaints against air pollution. Ambient air quality monitoring had been carried out since 1964, and the monthly average of sulphur oxides emissions sometimes exceeded 0.2 ppm. In 1970, 653 citizens were officially designated as patients suffering from pollution-related diseases.

Since 1966, the municipal government had conducted intensive monitoring by establishing monitoring stations at every square of 400 × 400 metres and had inspected all the 230 factories with smokestacks. Based on the data obtained from this study and simulation analysis, the municipal government determined how much each factory contributed to local air pollution and, using the result of the study, issued guidance to the factories on how they should deal with the pollution problem. In order to implement emergency pollution control measures, the municipal government organized the Special Guard, which consisted of 13 local officers, for Nishi Yodogawa Ward Pollution Control.

The target of the emergency measures was to attain, within two years, the national EQS for sulphur oxides, which stipulated 0.05 ppm as the annual average of the hourly value, whereas the national government had requested that the municipal government attain the EQS within ten years. The measures were commenced in 1970, and the ambient concentration, which had been 0.083 ppm in 1969, decreased to 0.042 ppm by 1972. The target of the municipal government was thus attained within two years after the commencement of the measures.[5]

Factors that contributed to successful administrative guidance

Avoidance of pollution control agreements

During the mid-1960s, the Yokohama Municipal Government succeeded in improving air quality by using pollution control agreements (PCAs) (Fujikura, 2005). The Yokohama Municipal Government concluded PCAs with local industries one after another whenever a company expanded its production facility or established a new facility. PCAs were widely appreciated as an effective tool for pollution control. In Osaka, opposition parties in the Osaka Municipal Assembly, the mass media and the local academics repeatedly demanded that the municipal government conclude PCAs with local industry. However, the municipal government refused to do so and continued to issue administrative guidance without concluding any PCAs with individual factories.

The municipal government countered that pollution control agreements bound not only factories but also the government as well. Once the municipal government reached an agreement with a factory, it would not be able to request any additional measures even when better technologies than those already subscribed in the agreements were developed. The municipal government was concerned that the pollution control agreements would become a sort of 'indulgence', allowing the factories to operate without adopting the 'best' measures being developed in anti-pollution technology, which was experiencing dynamic development. Moreover, PCAs concluded in Yokohama were not intended as instruments to control pollution from existing factories that had no plans to expand or renovate production facilities. Such factories had no incentive to conclude any agreements with local authorities. Thus, the Osaka Municipal Government opted to institute administrative guidance for factories instead of using PCAs.

Relationship between the government, business and citizens

Small industries in Osaka were technically and financially supported by the municipality and they felt that they were indebted to the municipal government. They felt something of an obligation to follow the guidance of the municipal government, and they invested in anti-pollution measures once they understood the pollution problem and the degree of their responsibility for it.

Company owners and factory managers often served as directors of local organizations such as LDAs. Many complaints against pollution were submitted to the municipal government through these local organizations, and the communities often requested that the municipal government hold meetings to inform the public about the situation of local pollution. Thus, since the directors of groups raising complaints oftentimes were also the polluters of the local environment, they were forced to adopt anti-pollution measures in order to hold their prestige in the community.

Once the municipal government accepted an individual plan for air pollution prevention from a factory, the government then requested citizens to endure the pollution until the measures were completed. Citizens trusted the municipal government and waited for the completion of the measures. Such mutual relationships among the municipal government, local businesses and citizens are what enabled the achievement of substantial pollution control in Osaka City.

Initiative of the mayor

Osaka City had been adversely affected by air pollution since long before the pre-war period. Dealing with the pollution problem had become a regular concern of the municipal government. Every mayor understood that the municipal government had to deal with the air pollution problem. Particularly since the 1960s, pollution control measures had been the most important consideration of the municipal government, and the mayor actively expanded environmental agencies related to air pollution control. In 1975, the number of staff involved in pollution control reached 192. Such initiatives by the mayors were welcomed by the local citizens.

Financial support

Among the various national and local measures for supporting the local business community,[6] there was a unique facility in Osaka for purchasing factories' land that particularly encouraged pollution control among the SMIs. A number of factories in Osaka City owned by SMIs were situated on plots that were too small to install any effective pollution control equipment; hence, the only viable options for them to address their pollution problems were either to relocate or shut down their business. According to one study by the municipal government in 1968, out of 290 factories that had caused (or were likely to cause) complaints from local citizens, 182 factories considered relocation. Under Osaka's facility for purchasing land, the municipal government bought the lots vacated after SMIs' factories relocated to other areas. By 1991, the municipal government had bought a total area of 192,780 square metres from 72 factories.

Through this facility, SMIs were able to obtain funds necessary for acquiring new land to relocate their business and also sometimes even new production facilities. This was because the municipal government bought their land in the city at the market price, which was much higher than the price of the new land where they relocated. Sometimes, the municipal government also assessed the environmental impact of the factory at the new site instead of having the assessment done by the local government responsible for the site; this was because the local governments trusted the capability of the municipal government in pollution control.

Through this, the municipal government was able to acquire land in the city, which it made available for establishing municipal parks and community centres. While factories utilized this facility, Osaka City eventually lost a significant portion of its local manufacturing industry. At the same time, nonetheless, pollution problems due to SMIs were solved.

Administrative guidance after the 1970s

Looking at sulphur oxide emissions currently, stringent national emission controls are enforced for stationary emission sources, and ambient air quality meets the EQS throughout Japan, including Osaka City. It has become unnecessary for the Osaka Municipal Government to issue administrative guidance to control sulphur oxide emissions. Presently, the municipal government issues administrative guidance to control nitrogen oxide emissions, dioxin emissions and offensive odours from stationary sources.

Most large Japanese cities have established municipal ordinances for pollution control, and the municipalities are exempt from prefectural pollution control ordinances. As an exception to the above, Osaka Municipal Government has not established a municipal ordinance for industrial pollution control, and Osaka Prefecture's ordinance is still applied to the emission sources within Osaka city. Because the prefectural ordinance stipulates no emission standards for nitrogen oxides and dioxins, the municipal government established municipal guidelines for these pollutants and issues administrative guidance based on these guidelines. The guidelines were neither based on any local ordinances nor approved by the municipal assembly. They are mere internal criteria adopted within the local administration. The municipal government regards the establishment of municipal ordinances for pollution control as still unnecessary because administrative guidance is effective enough to manage the local environment. The municipal assembly has not requested to enact a relevant municipal ordinance.

Large stationary sources of pollution are subject to the national emission control for nitrogen oxides, the goal of which is to attain the EQS. For smaller emission sources, which are not subject to the national control, administrative guidance is issued to cut the individual emission by approximately 40 per cent from the original amount of the emission. The guidance is usually issued when a production facility is newly built or replaced. Guidance is also issued during site inspections conducted by municipal officials. Usually, oral guidance is given at first, and then an official letter is provided at the next stage if necessary. An official letter is sometimes requested by a factory because it is necessary for the company's internal processing of the guidance. Only in cases where the environmental impact of a planned facility is deemed significant, the municipal government may still request a factory to submit an individual plan for pollution prevention as before.

In the cooperative atmosphere between the municipal government and local factories, local factories have followed the guidance without obvious resistance, and the municipal government does not demand stiff adherence to anti-pollution measures. No lawsuit has been initiated by the local industries concerning the legality of the guidance.

In the adoption of anti-pollution measures by SMIs, manufacturers of pollution control equipment play an important role. Many of the SMIs do not even have knowledge concerning their own production facilities. They do not even know whether they have registered their production facilities with the prefectural government according to the prefectural ordinance when they installed the facilities. Therefore, they have to entrust the planning and building of their facilities entirely to the manufacturers. The manufacturers enquire with the municipal government concerning the appropriateness of the pollution control equipment on behalf of their client before actually installing the equipment. If necessary, guidance is provided to the manufacturer instead of to their client.

The municipal government still regards PCAs with local factories as being unnecessary. After the Municipal Ordinance for Environmental Impact Assessment was established in 1998, the significance of the PCA was totally diminished. Now, monitoring and information disclosure can be enforced by the ordinance even more effectively than before.

2. Kitakyushu City

Kitakyushu's major industries were iron and steel, chemical products and stone and clay products. As the city resumed its industrial activities after the Second World War, Kitakyushu became covered in coal dust just as in the pre-war period.[7] As industrial fuel in Kitakyushu City was switched from coal to petroleum and consumption of coal decreased, the coal dust problem was naturally solved. However, sulphur oxide concentrations in ambient air gradually increased as petroleum consumption increased. The Fukuoka Prefectural Government was authorized by the Soot and Smoke Control Law to enforce air pollution control measures. It, however, had only a few officials responsible for pollution control and all of them were stationed in Fukuoka City, about 60 kilometres from Kitakyushu City. As a result, the prefectural government took almost no effective action in Kitakyushu.

Silent citizens

Despite severe pollution, the local people did not react, or they may have even accepted the problem. One of the reasons for Kitakyushu's silent citizens was their unawareness of the extent of the health effects of pollution. The majority of the local citizens were factory labourers and their families, and they regarded smoke from the factories as a 'symbol of prosperity'. People were particularly proud of the smoke from the Yahata Works, Japan's first modern iron mill. They thought that pollution was an unavoidable by-product of successful economic development. Everyone knew that the factories provided their jobs and had significantly improved their standard of living. Moreover, factory labourers did not regard Kitakyushu as a place to live but as a place to work. Most of them had come from the rural areas. They tended to return to their home-town villages and farms after retirement. Thus, their living environment was of lesser importance compared to the economic benefits they derived from working in the factories.

Another reason for the citizens' silence was the dominance of the Yahata Works in local politics and economics. Before the Second World War, the Yahata Works was a national iron mill and did not have to pay taxes to the city. It voluntarily 'subsidized' local development projects such as waterworks, schools and fire stations (Yahata Works, 1980, pp. 608–17; Hayashi, 1971, pp. 51–60). After the War, the Yahata Works was privatized and became one of Nippon Steel Corporation's ironworks. However, the Yahata Works' influence remained unchanged from the pre-war period. Two retired Yahata Works staff members served as mayors of Yahata City before the city became part of the newly formed Kitakyushu City in 1963. There was a closer relationship between the municipal government and Yahata Works and other similar enterprises than there was between business and government in other Japanese cities. The majority of the residents were either employees of these industries or families of employees. Under such circumstances, the local citizens hesitated to do anything that might threaten or anger these industries. To them, the likely cost of taking anti-pollution actions clearly surpassed the benefit of a cleaner environment (Fujikura, 2001).

Women's activities for raising public awareness

As time passed, an increasing number of labourers brought up their children and retired in Kitakyushu. As Kitakyushu became a place to call 'home', a place to reside and not just to work, more people became concerned about its pollution. It was the women who were first to take action in dealing with the problem. In the mid-1960s, some of the local women in Kitakyushu organized groups to study local pollution problems. The polluting enterprises quickly learned about the women's activities and used whatever means they could to stop them. Many of the husbands of the group members worked at the polluting factories, and the companies threatened them with personnel transfers.

The women did not directly challenge these enterprises. Instead, they petitioned the municipal government to implement effective anti-pollution measures and backed up their petitions with the results from their pollution studies. They also reported these findings to the public. The women produced an eight-millimetre film about local industrial pollution to show to the public. Newspapers and television stations reported their activities. This stimulated a general awareness of the impact of local air pollution on human health and the living environment. In a survey conducted by the municipal government in 1971, citizens were asked to select 'three issues about which they felt serious concern in their daily life' out of 17 issues. 'Smog pollution', at 49.6 per cent, was the highest, followed by 'traffic accidents' (46.0 per cent) and 'social welfare' (30.6 per cent).

Criticism of the Mayor and fear of election defeat

Mr Gohei Tani, the mayor of Kitakyushu, was a conservative politician supported by both the conservative Liberal Democratic Party and local industry. When he was first elected in 1967, he thought that pollution was strictly a prefectural matter. However, as the general public became increasingly aware of pollution problems, the number of citizens' complaints to the municipal government about air pollution increased from an annual average of 61 complaints during the 1960s to a record of 179 in 1971. Mr Tani was criticized for 'the delay in pollution control' and 'loose anti-pollution regulations' by both the local media and municipal assembly members. From the late 1960s, Kitakyushu City was always unfavourably compared by the local media and municipal assembly members of opposition parties with other industrial cities governed by leftist politicians, such as Yokohama City, where the socialist mayor managed to regulate emissions from major polluting sources by concluding PCAs with them.

Pollution problems became the most important issue in the 1971 mayoralty election. Mr Tani's rival, who was supported by the Communist Party, criticized his pro-business policies and appealed for more stringent pollution control. Mr Tani was forced to counter this attack by proposing a more progressive policy on pollution control. It was only through an intensive political campaign by the business community in support of Mr Tani that he obtained 61 per cent of the vote and was re-elected. However, the number of votes for his rival was still more than three times greater than the number of local Communist Party supporters.

Both the severity of the pollution problems and the loss of popular support for conservative politicians finally forced Mr Tani to take drastic anti-pollution measures. Having just upgraded the eight-member Pollution Control Division in 1970 to a 22-member Pollution Control Department in 1971, he again upgraded the department to a 47-member Pollution Control Bureau. While the mayor and senior municipal officials tried to deal with criticisms from the media, these officials (mostly engineers) simultaneously devoted themselves to enforcing measures that they believed would be the most effective in controlling pollution.

The mayor's fear of election defeat was shared by the local industry magnates, who were his supporters. They gradually recognized that there was no alternative but to support the municipal government's increasingly progressive pollution control measures. If pollution worsened, a leftist would replace the conservative mayor of Kitakyushu and very likely would impose even stricter demands upon them. In other cities, leftist politicians supported by local environmental activists were replacing conservative mayors. In 1971, for example, despite unfavourable prospects going into the mayoral election, a socialist politician succeeded in replacing the conservative incumbent mayor of Kawasaki City, where, as in Kitakyushu, the air was also heavily polluted by local industries. Kitakyushu's local industry was well aware of the very stringent measures being implemented in leftist-controlled cities. The industries' head offices in Tokyo were also greatly concerned about the increasing number of leftist local governments. They strongly desired that the mayor of Kitakyushu remain a conservative. It would be much better for the local factories to adopt costly anti-pollution measures according to the request of the conservative mayor rather than to be 'tortured' by a new leftist mayor (Fujikura, 2001).

Role of solid scientific data

In 1970, the Fukuoka Prefectural and the Kitakyushu Municipal governments and the major industries of Kitakyushu established the Joint Committee on Air Pollution Prevention in Kitakyushu. This joint committee included representatives of the 30 factories emitting 97 per cent of the city's sulphur oxides. In order to meet the national environmental quality standard (EQS) for sulphur oxides by FY1973, the municipal government concluded PCAs on sulphur oxide emissions (First Agreement) with 54 factories simultaneously in 1972, based on agreements reached by the joint committee.

The PCAs consisted of seven articles, which were identical for all factories and prescribed the following five obligations: the factory shall (1) attain the target for emission reductions prescribed in the attached annex; (2) positively adopt pollution control technologies and continuously upgrade the pollution control plan included in the annex; (3) consult and obtain the consent of the municipal government in advance when the annex is amended or an additional emission source is installed; (4) positively follow administrative guidance issued by the municipal government, cooperate with every survey conducted by the municipal government, submit requested information to the municipal government and conduct other necessary measures; and (5) accept site inspection by municipal officials.[8] By concluding such a simple PCA, the concerns of the Osaka Municipal Government that PCAs would bind the municipal government were resolved. Factories had to adopt the best technologies as they were developed even after conclusion of the PCA, and the municipal government remained free to issue additional guidance to the factories.

An important issue was the fair distribution among the factories of the financial and operational burdens entailed in the anti-pollution measures. Ideally, the burden would be shared among these factories proportionate to their responsibility for the air pollution problems. However, they did not actually know the degree to which their operations were responsible for pollution. In order to solve this problem, the municipal government utilized the results of a CSIP carried out by MITI.

Prior to the CSIP conducted in Osaka in 1972, MITI started a CSIP in 1969 in the Kitakyushu area in order to forecast the environmental impact of a new industrial complex planned for Kitakyushu. It was Japan's first comprehensive environmental impact study. Data accumulated by the municipal government through continuous monitoring were utilized for this study.[9] Moreover, a fluorescent substance was sprayed from a helicopter in order to determine meteorological elements. Based on these data, a wind-tunnel experiment was conducted and each factory's share of responsibility for existing air pollution was determined.

Based on the CSIP experimental results, the municipal government recommended that each factory adopt specific measures, such as installing higher smokestacks. In the annex of each agreement, there was described a detailed plan prepared by each factory for reduction of sulphur oxides based on the city's recommendations. By utilizing the annex, the municipal government was able to provide each factory with ongoing individualized and comprehensive administrative guidance.

Once solid scientific findings were presented to require specific anti-pollution measures from the factories, the only way for industry to resist was to challenge that data. One company carried out the same wind-tunnel experiment at its own expense only to confirm that the city's data were correct. Because the data could not be disproved, local industry had no choice but to comply with the city's advice. After that, the factories paid more attention to the fairness of the financial and technical burdens placed on local industry for cutting emissions rather than the reliability of the scientific findings of the municipal government.

Such arguments concerning fairness were often facilitated under the initiative of Nippon Steel, the largest company operating in Kitakyushu and the traditional leader of the local industries. All local industries tacitly agreed that Nippon Steel would represent them and, as a result, the municipal government only needed to negotiate directly with the Yahata Works of Nippon Steel.

Once the municipal government and Yahata Works reached an agreement on any anti-pollution measures, the other companies would comply with the agreement. The social conscience of the Yahata Works facilitated this process. The management had always been socially elite and taken pride in their company's role in the local society. Once Yahata Works accepted the city's stringent anti-pollution measures as being necessary for the general welfare of the city, it persuaded the other enterprises to comply (Shikata, 1991).

By utilizing higher smokestacks, the EQS was attained by 1973. However, in 1973 the national government replaced the existing EQS for sulfur oxides with a new EQS for sulphur dioxide set at 0.04 ppm for the daily average one-hour value and at 0.1 ppm for the maximum one-hour value, almost three times more stringent than before. The national government required the municipal government in Kitakyushu to make the ambient air quality meet the new EQS within five years. The members of the joint committee concluded a new pollution control agreement (Second Agreement) to further reduce total emissions in Kitakyushu to one third in order to meet the new EQS by 1977.

In order to comply with the Second Agreement, factories had to switch permanently from high-sulphur to low-sulphur heavy oil. In the same manner as the previous agreement, the fair distribution of the financial burden among the factories was seriously taken into account. In 1973, the municipal government advised the factories to reduce their emissions uniformly by 30 per cent within a year. After the municipal government confirmed that all of the factories had cut emissions by 30 per cent, it advised them to further reduce emissions by 20 per cent. By 1976, the sulphur dioxide concentrations of ambient air at all the monitoring stations met the new EQS.

Local industries' incentives for concluding a PCA and adopting anti-pollution measures

There was another political reason for local industries to support the conservative mayor by following the advice given by the municipal government in spite of high economic cost. The annexes of the PCAs included trade secrets about raw materials, kinds of fuels, the sulphur content of these fuels, fuel consumption, maximum emissions, etc. Industries could trust the conservative municipal government to treat these annexes as classified information. They were concerned that these secrets might be disclosed to the public if leftist politicians replaced their conservative mayor. For example, all the contents of pollution control agreements concluded in leftist-controlled Yokohama City were opened to the public. In order to keep the annexes secret, it was imperative that the conservative mayor not be replaced by a leftist politician and that local industries follow the advice of the municipal government.

There were also technically reasonable aspects to the adoption of PCAs. When the ambient air was significantly polluted, the mayor issued a 'smog alarm' in accordance with national law. Once a smog alarm was issued, factories were required to reduce their operations or switch to a more expensive fuel containing lower sulphur. The alarm sometimes lasted several days if weather conditions remained stable, forcing the factories to pay a high cost during the entire period. It was cheaper for the factories to adopt anti-pollution measures. By switching to low-sulphur fuel before the issue of the alarm, the alarm could be avoided and the factories only had to use low-sulphur fuel for one day. By concluding a PCA, a factory was able to obtain local weather information from the municipal government[10] and was then able to take measures before the situation worsened enough to warrant a smog alarm.

Another technical advantage of concluding a PCA for industries is that it shortens the period required for administrative procedures. According to the Air Pollution Control Law, every factory is required to submit advance notice to the authorities when it establishes a new emission source. Ordinarily, the factory has to wait 60 days until the report is officially confirmed. However, the period is shortened to 20 days for factories that have concluded a PCA. The annexes of the PCA include the industry's production plan for the coming five years, with an estimation of emissions during that period. Because the factory and the municipal government have agreed on the plan, the factory can implement the production plan without intervention from the municipal government. Emissions from subcontractors' factories located within the grounds of Nippon Steel can be included in the emissions of Nippon Steel, and the factories are exempted from submission of a plan.

Anti-pollution measures

As industries were required to adopt pollution control measures throughout Japan, the demand for low-sulphur heavy oil increased and the supply became insufficient. The price of low-sulphur heavy oil relative to high-sulphur heavy oil increased dramatically. In 1969, whereas one kilolitre of high-sulphur heavy oil cost 5,900 yen, one kilolitre of Indonesian heavy oil containing only 0.3 per cent sulphur was 6,100 yen. The price difference amounted to only 3 per cent; by 1973, however, this difference had jumped to 61 per cent (Yahata Works, 1982).

Both the high cost and the difficulties of procuring low-sulphur fuel had inconvenienced Japanese industries. For example, when the municipal government required a particular factory to reduce its sulfur dioxide emissions to one sixth between 1970 and 1975, the factory estimated that the additional expense of switching to low-sulphur oil would eliminate half of its ordinary profits. Despite strong opposition within the company, it nonetheless accepted the cost increase and changed fuels.

Desulphurization technology had become almost practical by 1974, but the municipal government remained disinterested in flue gas desulphurizers. When a factory prepared a plan to install desulphurization equipment, the municipal government not only requested the installation of back-up emergency equipment, but also asked for a commitment to stop the whole operation whenever the desulphurization equipment stopped functioning. Thus, with the exception of factories such as steelworks that used coal as a raw material and were already equipped with desulphurization technologies, the switch to low-sulphur fuel was the only option for Kitakyushu industries to meet the city's anti-pollution guidelines.

The switch to low-sulphur fuels and the energy-saving measures implemented after the two oil shocks in 1973 and 1978 had substantially reduced sulphur dioxide emissions in Kitakyushu. For example, the actual sulphur oxides emissions of the Yahata Works in 1990 were estimated to be about 2.2 per cent of the 1970 emissions, given the same amount of production. The switch to low-sulphur fuels and changes in raw materials accounted for 42 per cent, the majority, of the emission reductions. Energy savings and recycling accounted for 33 per cent and flue gas desulphurization, 25 per cent (IES, 1996).

Pollution control agreements after the 1970s

Kitakyushu's PCAs were authorized by the Kitakyushu Pollution Control Ordinance, established in 1970. All factories in Kitakyushu City were required to conclude a PCA with the municipal government. All of the draft PCAs were examined by the Municipal Pollution Control Committee, which consisted of external experts, before the municipal government officially concluded the agreements. As the number of factories increased in the 1980s, however, the administrative cost became significant. The municipal government amended the ordinance so that PCAs were only concluded with factories having production facilities of a certain size. Smaller factories were only required to submit a covenant to the municipal government.

The original objective of the PCAs was to attain the EQS for sulphur dioxide. Once that objective was attained, the rationale of the PCA diminished. After total emission control for sulphur oxides was introduced by the national government in 1974, it became possible to conserve the atmospheric environment without the PCAs. Other pollutants could also be controlled without PCAs. Acceptance of site inspection and information disclosure was assured by the introduction of instruments other than PCAs, such as the Municipal Ordinance on Environmental Impact Assessment.

Administrative guidance is currently being issued based neither on a PCA nor on a covenant. It is issued during the implementation of national and local regulations. The PCAs have become mere instruments for publicizing the city's efforts towards environmental management. One example of a PCA as a publicity instrument was seen when the municipal government decided to establish a facility to decompose polychlorinated biphenyls (PCBs) in the city in 2001 in accordance with the national policy. In order to obtain public acceptance, the municipal government convinced the citizens that the facility would be properly managed by pointing to a PCA concluded with the facility.

Administrative guidance is also issued when complaints are submitted to the municipal government from citizens. Most complaints are concerning noise from factories. In such cases, factories seldom are violating any regulations but the municipal government advises them of the technical and financial measures to implement. If necessary, a soft loan is provided by the government. Even if the pollution (including noise) is not improved to a satisfactory level, citizens compromise when they see the local officials frequently visiting the site and working hard. Citizens continue to trust the municipal officials.

3.   Factors contributing to successful administrative guidance

Although the administrative guidance issued both by the Osaka and Kitakyushu municipal governments did not prescribe sanctions against non-compliance, it did successfully reduced sulphur oxide emissions during the early 1970s. In both cities, there are several common factors that contributed to the effectiveness of the guidance (Fujikura, 2002).

Complaints of citizens

During the 1960s and 1970s, confrontational anti-pollution movements, which included demonstrations, occupation of polluting factories and sit-ins, occurred in many Japanese cities. Osaka City was no exception. However, the Osaka Municipal Government believed that these radical activities in the city were mainly organized by leftist political groups as part of their anti-government campaign, and Osaka City found that very few of the participants were actually residents who were affected adversely by pollution. The municipal government concentrated on responding to complaints submitted through the local communities, and it was these complaints that actually forced local factories to adopt anti-pollution measures.

In Kitakyushu, there were no confrontational movements. Under the politically suppressive conditions, citizens could express their discontent only by voting for the rival of the incumbent mayor. However, this was effective enough to force the mayor and senior municipal officials to advertise their efforts for pollution control through the mass media. The mayor had to recover support from citizens.

The citizens' expression of discontent was sufficient to force municipal governments to adopt substantial anti-pollution measures even in the absence of confrontational citizens' movements. On the other hand, the citizens trusted the local officials in spite of the mass media's accusations concerning municipal governments. In both cities, municipal officials actually devoted themselves to their duties towards improvement of the local environment, and citizens recognized their efforts. As a result, the mayor of Osaka was able to maintain popular support, and the number of votes for the incumbent mayor of Kitakyushu increased after the local environment improved.

Communication between the municipal governments and local industries

In both cities, the municipal governments maintained close connections with local industries. In Osaka, local officials held and/or attended numerous meetings with local industries and explained the necessity of anti-pollution measures. In Kitakyushu, the Joint Committee on Air Pollution Prevention played a central role in discussions on the necessity and effectiveness of the measures and in facilitation of adoption of the measures by the local industries in spite of high cost. Neither of the municipal governments compelled the industries to follow their policy, but once each factory was convinced of the necessity of adopting the measures, the factories made a sincere effort to implement them.

Leadership of the mayors and high capability of the municipal governments

Mayors hold considerable discretionary policy-making power in governing their municipalities. Local governments in Japan are relatively independent of the national government, and the heads of the local governments sometimes adopt local policies that actually contradict national policies. Within the local government, the heads control both the budgetary and personnel affairs of the local government, unlike in the national government, where most of the decisions are made through the bottom-up approach. During the 1970s, local government had sufficient revenue as the economy was rapidly developing, and the mayor of Osaka was able to allocate more than 10 per cent of the municipal budget at his own discretion. Kitakyushu and other industrial Japanese cities were also in a similar budgetary condition. With such ample budgets, mayors' decisions were easily materialized.[11] The governors and the mayors are elected by direct local election, and they are sensitive to the local opinions. Once the mayors knew of local people's discontent and decided to reduce pollution, the cities rapidly strengthened the relevant personnel and budget and developed the capacity to provide effective administrative guidance.

Municipal officials are capable. In Japan, local governments can recruit capable officials, particularly engineers, because local officials enjoy a high social standing and receive a relatively high salary and tenure. No cases of corruption involving pollution control were reported. The municipal officials accumulated data on science and technology in order to build their capacity to convince engineers of large industries and to advise SMIs. When large industries resisted adopting the measures proposed by the municipal officials by insisting that the measures were technically impossible, the municipal officials studied measures already adopted at other factories in the same line of business located in other areas and were able to refute them.

Scientific data

Both in Osaka and Kitakyushu, the CSIP conducted by MITI was a crucial piece of evidence that the municipal officials used as solid scientific data to convince local factories. By showing the actual contribution of each smoke-stack to the local air pollution, the officials convinced the factories of the necessity of anti-pollution measures. Because the administrative guidance had no solid legal basis, hard scientific data was the only means available to the municipal government to force factories to adopt costly anti-pollution measures.

Both municipalities recognized the importance of scientific data and started environmental monitoring in the early period of industrialization in Japan. The predecessor of the Osaka Municipal Institute of Environmental Studies, Osaka Institute of Sanitation, was established in 1906, and the Osaka Research Committee on Soot and Smoke Prevention organized by the municipal government started monitoring coal dust fall as early as 1913. Kitakyushu City was established in 1963 by merging five cities, and one of the five cities, Tobata City, had already started monitoring dust fall in 1956. Kitakyushu Municipal Institute of Sanitation was established in 1965.

Availability of technologies and funds

In Kitakyushu, the major polluters were large industries that had advanced technical knowledge and adequate funds. In particular, Yahata Works developed clean production technology for steel production and succeeded in significantly decreasing the amount of its sulphur oxide emissions.

In Osaka, the major polluters were SMIs and the heating of buildings in commercial areas. They did not possess knowledge of anti-pollution technologies, so the municipal officials frequently visited exhibitions of new pollution control technologies and brought the brochures obtained at the exhibitions to meetings with local industries. In addition to the above-mentioned financial support by purchasing land, the municipal government established several financial support mechanisms for the local SMIs. The municipal government requested the national government to establish new tax measures to encourage the SMIs to adopt anti-pollution measures.[12]

Manufacturers of pollution abatement equipment also played an important role in pollution reduction by SMIs. As the SMIs often had little knowledge about pollution control and left anti-pollution measures entirely to the manufacturers, the manufacturers examined the adequacy of the technology on behalf of their clients.

Incentives for compliance

As mentioned above, there were several reasons for local industries to follow administrative guidance despite the high cost and despite the fact that there were no sanctions for non-compliance. One fundamental reason for dealing with the complaints and discontent of citizens was that, if they failed to do so, their local prestige would be tarnished and their business would be hindered. Industries feared that they would be reported as environmental polluters by the mass media, and they regarded harmony with the local community as important. Large industries in particular have the traditional idea that they should contribute to local communities. In large industries, once a company decided to adopt a pollution control policy, actual emission levels became a subject of competition among the factories of the same company. Employees made efforts to reduce pollution in order to win the race for promotion within a company.

4.   Evaluation of administrative guidance

Effectiveness

Voluntary approaches to environmental management have been widely implemented in developed nations as supplements or alternatives to environmental legislation. The OECD identified three main types of voluntary approaches: public voluntary programmes, negotiated agreements including PCAs and unilateral commitments (OECD, 1999). Then, the OECD evaluated them through case studies in member countries (OECD, 2003, p. 14), and concluded as follows: while the environmental targets of most – but not all – voluntary approaches seem to have been met, the environmental effectiveness of voluntary approaches is questionable. How much environmental improvement was really attained by the voluntary approaches was unclear.

While the OECD studied Japanese PCAs and concluded that the PCAs 'are playing a prominent role in Japanese environmental policy' (OECD, 1999, p. 71), it did not analyse the administrative guidance issued in Japan. Administrative guidance in Osaka and Kitakyushu had an even more 'voluntary' nature than the voluntary approaches analysed in the OECD studies. OECD mentioned that '[t]he public authority commitment generally consists of not introducing a new piece of legislation (e.g., a compulsory environmental standard or an environmental tax) unless the voluntary action fails to meet the agreed target' (OECD, 2003, p. 19). In both the cases of Osaka and Kitakyushu, however, no penalties were prescribed in the event that the targets of the guidance were not met, and there were no provisions to compel the factories to comply with the guidance. Furthermore, the guidance was often given only orally. Guidance in Osaka in particular had no official basis at all.

At variance with the OECD's conclusion, the administrative guidance issued in both cities was, in fact, environmentally effective. Based on the result of scientific analysis, both of the municipal governments recognized that it was *impossible* to attain the EQS within the period specified by the national government without introducing additional measures, and so they issued administrative guidance; the administrative guidance was not supplemental measures but necessary measures for attaining the environmental target.

Efficiency

As for the efficiency of the voluntary approaches, OECD concluded that 'administrative and transaction costs vary greatly' (OECD, 2003, p. 15), and '[t]he *economic efficiency* of voluntary approaches is *generally low* – as they seldom incorporate mechanisms to equalize marginal abatement costs between all producers, *inter alia* because environmental targets tend to be set for individual firms or sectors, rather than at a national level' (OECD, 2003, pp. 14).

The administrative cost of the guidance seems to be high. While relevant data on the cost during the 1960s and 1970s is not available, administrative guidance in Osaka nowadays is as follows. In 2004 in Osaka City, there were 3,238 emission sources as designated by national laws regarding air pollution or Osaka Prefecture Ordinance. The frequency of the site inspections at each site is prescribed by the national laws or the ordinance, and 1,501 site inspections were conducted in 2004. One site inspection usually required approximately two or three hours, and a municipal inspection team visited three sites a day. Administrative guidance was issued mostly at these opportunities. Eight municipal officials specialize in inspections and administrative guidance involving air pollution, and 111 engineers at municipal health and welfare centres located in every one of the 26 administrative wards are additionally involved in these duties.[13] The cost of this is high, and it appears that a substantial reduction of the pollution caused by numerous small emission sources without incurring a high administrative cost would be otherwise difficult.

Administrative cost in Kitakyushu seems to be lower than in Osaka. This is due to the small number of the emission sources in Kitakyushu. In Kitakyushu, during the 1970s, the number of the major emission sources was less than 60, and all of them concluded the same PCAs with the municipal government. Moreover, Nippon Steel, the local leader of industry, facilitated processes leading to agreements between the municipal government and other local factories. This definitely lowered the administrative cost. However, as the number of the factories increased, the cost of concluding PCAs became significant, and so the municipal government finally halted PCAs with small factories and requested them to submit covenants instead.

Although hard evidence is unavailable, the cost of pollution control at each factory during the 1970s seems to have been increased by administrative guidance. The guidance often required that the factories implement measures in addition to the ones required by national legislation.[14] Otherwise, the target (the EQS) would not have been attained. Like the cases studied by the OECD, environmental targets were set for individual factories both in Osaka and Kitakyushu, and the administrative guidance did not provide for equalization of marginal abatement cost. This was because fair apportionment of the emission reduction among the local industries was more important to them than the equalization of marginal abatement costs. The individual target had to be, therefore, based on solid scientific data. Moreover, Japanese citizens, the mass media and, as a result, the government have sought the 'best available' technology, regardless of its economic cost. It was difficult for the factories to choose a less environmentally effective measure even though it was environmentally adequate and more economical. This economic inefficiency can partly be attributed to the emotional circumstances.

Transparency and participation

Administrative guidance is closed and opaque. Both in Osaka and Kitakyushu, stakeholders other than the polluting factories and the municipal governments, such as victims of the pollution, had no opportunity to participate in the process of preparing the pollution control programme and no access the information of individual factories. For example, in Osaka, the plan for air pollution prevention was submitted to the municipal government and, because the municipal government regarded it as totally voluntary despite the advice provided by municipal officials during the preparation, the municipal government did not issue any official acknowledgement of the plan. This implies that no detailed plan from any individual factory officially existed in the municipal government office, and so ordinary citizens were unable to access the document. In Kitakyushu, a substantial pollution control target and plan were prescribed in the annex of the PCA, which was not open to the public.

During the 1970s, stakeholder participation was not on the agenda of the national or the local governments. Many of Japan's local governments considered participation of ordinary citizens in local policy-making to be unnecessary because the local governments held a sufficient number of meetings to inform the citizens of their policy and adequately communicated with them. Influential NGOs demanding stringent environmental policies had not yet emerged, and the environmental movements were mostly conducted by the local communities that were affected. Furthermore, a matter of concern of the ordinary citizens, even of the members of the affected local communities, was not participation in local environmental policy-making but actual improvement of their local environment. They actually trusted the local officials. Because the administrative guidance was based on solid scientific data and local officials were conscious of their responsibility, the guidance had less arbitrariness. Therefore, the opacity of the guidance and non-participation of citizens in the formulation of the guidance did not become an issue.

Implications for technical cooperation with developing countries

Administrative guidance to every factory is one possible approach to pollution control in developing countries, particularly in the areas where a large numbers of SMIs are causing pollution, as they were in Osaka City. In order to examine the applicability of administrative guidance in developing countries, the social background in Osaka and Kitakyushu are summarized below:

1. Local people feel discontented with pollution and they have the means to express their discontent, such as submitting complaints to the municipal government or voting for the opposition party in the local election.
2. Heads of a local government are directly and freely elected by the local people, or the heads are, at minimum, sensitive to local opinion.
3. Local government is sufficiently capable of collecting solid scientific data and convincing local factories to adopt anti-pollution measures.

4. Local officials are free from corruption, are conscious of their roles and are trusted by the local people.
5. The central government is capable and reliable if the authority for local environmental management is not delegated to the local governments.
6. Companies that are polluters have some kind of incentive to reduce their pollution, whatever their objectives may be.

In Japanese industrial cities, including Osaka and Kitakyushu, pollution occurred because of a number of factors, such as lack of adequate national legislation, lack of public awareness and lack of technologies (Fujikura, 2005). In both cities, however, the above-mentioned conditions were fortunately satisfied and administrative guidance became effective once it started. Most pollution problems nowadays are most likely happening in areas that are lacking some of the above-mentioned conditions. If all of the conditions are satisfied, pollution is likely to be improved. In order to make administrative guidance effective, effort should be made to improve the conditions of the targeted areas.

It is difficult for foreign donors to intervene in the administrative structure of developing countries, even though this is regarded as crucial for pollution control. The donors should concentrate on enhancing the capacity of the local (or central) government. In addition to providing support to the local government, the role of local manufacturers of pollution control equipment should also be recognized. When the local market for pollution control equipment becomes large enough and the manufacturers have adequate experience and technology, they can be expected to play a certain role as technical consultants to SMIs, as seen in Osaka City. The capacity of the local governments for pollution control and the local market for pollution control equipment will develop in tandem. Implementation of pollution control measures increases the demand for pollution control equipment, and this stimulates the development of local manufacturers. Support for the development of the local market can be also an option of the donors; for example, providing knowledge and funds for the developing countries to promote soft loans for the SMIs, and/or promoting cooperation between the pollution control equipment manufacturers in developed and developing countries may be effective.

The feasibility of introducing PCAs in developing countries should be further examined. PCAs have actually been playing a significant role in local environmental management in Japan. Particularly during the early 1970s, usage of PCAs was crucial for the implementation of urgent and necessary measures in some industrial cities because the municipal governments were not adequately authorized by the national government with the power to control polluting sources. However, as seen in Kitakyushu, once national and local regulations were fully developed, the rationale for the PCA diminished. The major objective of many PCAs currently being concluded in Japan seems to be merely to obtain public acceptance of new facilities, such as waste treatment facilities. In many developing countries, legislation for environmental management has already been developed, but the enforcement of the regulation is often insufficient. Introducing PCAs without developing the capability of the local governments may not be effective and may only increase administrative cost.

Acknowledgements

This study is primarily based on a number of interviews with former and current senior municipal officials. The author would like to express special thanks to the following interviewees: Mr Takeshi Matsumiya, Mr Kazuhiko Takesada, Mr Kazuhiro Oishi, Mr Kazuo Inoue and Dr Isao Yamamoto of the Osaka Municipal Government and Mr Satoshi Nakazono of the Kitakyushu Municipal Government.

Notes

1. The Soot and Smoke Control Law was established as a compromise between the Ministry of Health and Welfare and MITI. The former had concerns over health damage caused by the air pollution and the latter, representing the interests of industry, had concerns over the negative impact of pollution control on industrialization. The authority to implement the law was not delegated to municipal governments but to prefectural governments. Emission standards were applied only within designated 'pollution-control areas' and, thus, no regulations were enforced outside these areas. The emission standard for sulphur oxides (sulphur dioxide and sulphur trioxide) was set at 2,200 ppm. This standard could easily be met without industries having to undertake additional pollution control measures. Power plants, which were the largest contributors to air pollution, were not subject to the law. Legal enforcement, particularly for prevention of air pollution, could not be expected.
2. In 1969, the Cabinet adopted the following EQS (the first one) for sulphur oxides:
   a) the number of hourly values less than 0.2 ppm shall exceed 99 per cent of the total number of observed hourly values throughout the year;
   b) the number of days when the daily average of the hourly values is less than 0.05 ppm shall exceed 70 per cent of the total number of observed days throughout the year;
   c) the number of hourly values less than 0.1 ppm shall exceed 88 per cent of the total number of observed hourly values throughout the year; and
   d) the annual average of the hourly value shall not exceed 0.05 ppm throughout the year.

   Upon the establishment of the EQS, the National Cabinet grouped designated geographical areas into four categories and set a target year for the attainment of the EQS in each area. The EQS for sulphur oxides had to be attained within ten years in large industrial cities, including Osaka and Kitakyushu, within five years in mid-size industrial cities and as soon as possible in other areas.
3. The oil refinery industry obtained financial support from MITI for its own R&D on heavy oil desulphurization technologies and the dissemination of them. As a result, the average sulphur content of heavy oil decreased from more than 2.5 per cent in 1967 to 1.5 per cent in 1972, and further decreased to 1.3 per cent in the

late 1970s. MITI distributed low-sulphur heavy oil to all polluted areas designated by the Cabinet when defining the target year for attaining the EQS for sulfur oxides. These efforts substantially contributed to the reduction of sulfur oxide emissions throughout Japan starting in the mid-1970s (Terao, 1994).

4. As the national government recognized that the Soot and Smoke Control Law of 1962 had failed to control sulphur oxide emissions from factories and nitrogen oxides from automobiles, it established the Air Pollution Control Law in 1968 to replace the 1962 law. The major differences between the new and old laws were as follows: in the new law, a) area designation was abolished and, thus, all areas became subject to control; b) toxic substances such as cadmium and chlorine were designated as air pollutants; c) a direct punishment provision was introduced, making polluters punishable without the need for any prior order to adopt anti-pollution measures; and d) prefectural governors were allowed to introduce ordinances with more stringent regulations than national ones, except for the emission standards for sulphur oxides.

   With regard to sulphur oxides, the uniform emission standard for sulphur oxides concentration in flue gas was abolished and new emission standards were established based on the location and the smokestack height of emission sources.

5. The Nishi Yodogawa area has also been polluted by nitrogen oxides emitted from vehicles driving through a national road intersection in this area. This pollution problem has not been solved yet.

6. From the 1960s, the national government provided economic incentives to facilitate factories' adoption of anti-pollution and energy savings measures. These incentives included soft loans provided by public financial institutions and local governments, as well as tax exemptions. For the SMIs, 40 billion yen was provided by public financial institutions in 1975, and another 40 billion yen was provided by local governments in the same year. Decreases in tax revenues of the national government, which were due to tax reduction measures intended to promote pollution control investments, reached 100 billion yen in 1975. Of this decrease, reductions in the direct national tax accounted for 60 billion yen, which was equivalent to 20 per cent of the total reduction in corporation taxes for the year (Terao, 1994).

7. The residential district of Shiroyama, which was surrounded by industrial areas, continuously was listed as Japan's worst site of dust fall from the late 1950s to the mid-1960s. A monthly record of 123.8 tons of dust fall per square kilometre was recorded in September 1966. Roofs caved in under the weight of the dust. Oily dust stuck to electric wires formed 'dust icicles'. Every elementary school classroom was equipped with two air-cleaning machines. Sprinklers were installed in the playgrounds to prevent the dust from blowing into the classrooms. Pupils had to gargle every day using special 'gargling equipment'. Despite the fact that Kitakyushu schools took every measure they could for adequate health management, large numbers of pupils continued to suffer from pollution-related illness. The proportion of absentees in Tobata Ward schools during 1963–5 always exceeded 10 per cent and sometimes even exceeded 50 per cent (Hayashi, 1971, pp. 111–18).

8. The PCAs prescribed factory site inspections to be performed by the municipal officials. Site inspections were usually conducted once or twice a year before winter; weather conditions in winter often exacerbate air pollution. The major objective of the inspections was for the local officials to identify points that the factories needed to check in the case of a smog alarm. The factories were notified of the inspections in advance. There were concerns that advance notification would hinder proper inspection activities. However, the cost of switching fuels and/or

turning on pollution control equipment for the short duration of the site inspection in order to trick municipal officials was also significant, and such activities did not necessarily contribute to the cost reduction as much as appropriate pollution controls did. Mass-media reports convinced the citizens of the effort of the municipal government and factories. The factories were not reluctant to accept the inspections as long as other factories were equally inspected in the same manner.

9. Monitoring began in 1959 of the dust fall, sulphur dioxide and iron oxides in the ambient atmosphere in the entire Kitakyushu area.

10. The municipal government established an observatory on a mountaintop in the city in order to forecast air pollution.

11. Presently, the national government and most of the local governments in Japan face serious budgetary deficits. The mayors and the governors can allocate only a few per cent or less of their local budgets to implement their policies. It is much more difficult now than during the 1970s for the heads of the local government to implement their initiatives.

12. Among the various tax measures for pollution control, reduction of the term of depreciation was effective, particularly for SMIs. As it was difficult for SMIs to take a long-term perspective of their business, they preferred to complete the payment of their property tax as soon as possible while business was profitable. The Osaka Municipal Government requested that the national government establish this tax measure to reduce the term of depreciation, and the measure was highly appreciated by SMIs, although it did not reduce the overall amount of taxes they had to pay.

13. In accordance with the Sewerage Law and the Osaka Municipal Ordinance for Sewerage, the municipal government conducted 8,600 site inspections a year at approximately 3,000 factories. One site inspection takes between 20 and 40 minutes, and 14 municipal officials are devoted to the task.

14. Adopting additional measures does not always increase the cost. Currently, the Osaka Municipal Government provides guidance on the installation of low NOx burners, which are approximately 20 per cent more expensive than ordinary burners. However, the incremental cost is often soaked up in the total cost of the production facility. Due to competition, it became difficult for the manufacturer of the production facility to simply add the cost to the selling price.

    Moreover, there are cases where the administrative guidance of the Osaka Municipal Government decreased the water pollution control cost. There are some manufacturers who attempt to sell a wastewater treatment facility with an unnecessarily large capacity. In such cases, the municipal government provides technical advice to the factory on appropriate measures, i.e. smaller equipment. This guidance actually reduces the cost.

References

Fujikura, Ryo (2001) 'A Non-confrontational Approach to Socially Responsible Air Pollution Control: The Electoral Experience of Kitakyushu', *Local Environment*, 6(4), 469–82.

Fujikura, Ryo (2002). 'Nihon no chihou-kokyodantai no Iou-sankabutsu-taisaku: kodo- keizai-seichoki ni jissisareta kogai-boshi-kyotei to gyosei-shido' (Local Measures in Japan for Controlling Sulphur Oxide Emissions: Pollution Control Agreements and Administrative Guidance Adopted during the Era of Rapid Economic Development), in *'Kaihatsu to kankyo' no seisaku-katei to dainamizumu: nihon no keiken, higashi-ajia no kadai* (*Dynamism of the Policy Process in 'Development and*

Environment': The Experiences of Japan and Problems in East Asia). T. Tadayoshi and K. Otsuka (eds), *Chiba,* Institute of Development Economics (in Japanese).

Fujikura, Ryo (2005) 'The Background to Successful Air Pollution Control in Japan', in Wilfrido Cruz, Adoriana Bianchi and Masahisa Nakamura (eds), *Local Approaches to Environmental Compliance: Japanese Case Studies and Lessons for Developing Countries,* Washington, DC, World Bank Institute, 19–51.

Hayashi, Eidai (1971) *Yahata no kogai (Pollution in Yahata),* Tokyo, Asahi Shimbun.

IES (Institute of Environmental Systems) (1996) *Kankyo kosuto to sangyo, kigyou (Environmental Expenditure and Industry),* IES Report No. 5, Fukuoka, IES of Kyushu University, 12 (in Japanese).

OECD (1999) *Voluntary Approaches for Environmental Policy – An Assessment,* Paris, OECD.

OECD (2003) *Voluntary Approaches for Environmental Policy – Effectiveness, Efficiency and Usage in Policy Mixes,* Paris, OECD.

Osaka Municipal Government (1994) *Osakashi kogai taisakushi (History of Pollution Control of Osaka City),* Osaka, Global Environmental Centre (in Japanese).

Oyama, Kosuke (1996) *Gyoseishido no seijikeizaigaku (Political-economics of Administrative Guidance),* Tokyo, Yuhikaku (in Japanese).

Shikata, Hiroshi (1991) *Kemuri wo hoshi ni kaeta machi (A City Turned Smoke into Stars),* Tokyo, Kodansha, 40–5 (in Japanese).

Shindo, Muneyuki (1992) *Gyoseishido (Administrative Guidance),* Tokyo, Iwanami Shoten (in Japanese).

Terao, Tadayoshi (1994) '*Nihon no sangyo seisaku to sangyo kogai*' (Japanese Industrial Policy and Industrial Pollution), in: R. Kojima and N. Fujisaki (eds) *Kaihatsu to kankyo: ajia 'sinseichoken' no kadai (Development and the Environment: Problems in New Industrializing Asia),* Vol. 4, Tokyo, Institute of Development Economics (in Japanese).

Yahata Works (1980) *Yahata seitetsusho 80 nen shi – bumon shi gekan (an Eighty-Year History of the Yahata Works – a History of the Departments)* Kitakyushu, Nippon Steel Corporation (in Japanese).

Yahata Works (1982) *Enerugi bu 60 nen no ayumi (A Sixty-Year History of the Energy Division),* Kitakyushu, Nippon Steel Corporation, 486–7 (in Japanese).

Part II
Dynamism of the Environmental Policy Process in Asia

5

Environmental Policy Planning under Imperfect Market and Government Capacity: A Case of Air Pollution Abatement in China

Nobuhiro Horii

Introduction

It is well known that China suffers from severe air pollution. China, however, has also made great progress in addressing the problem. For instance, statistics show that emissions of SO_2 and smoke dust, which are two main polluting substances, have substantially been reduced from 23.7 million tons and 17.4 millions, respectively, in 1995 to 19.2 million tons and 10.1 million tons in 2002. This represents a reduction of 14.1 per cent and 41.9 per cent, respectively. This is a great achievement for China, considering the high growth rate of the Chinese economy maintained during that same period and the accompanying increase in energy consumption, which is the source of pollution.[1] Such achievements would not have been possible without policy measures known as command and control measures. Above all, China has managed to implement policy measures that are imperative for pollution reduction. Specifically, China has implemented the compulsory closure of small coal mines and small power plants, as well as the mandatory shutdown of small enterprises in such high-energy-consumption industries as the steel and cement industries.

On the other hand, despite the above-mentioned achievements, China still tops the world in terms of SO_2 emission, and it is necessary for China to further reduce pollutants in the future. However, since the actions that are easy to take and can cause significant impact have already been taken, the cost effectiveness of the future measures will be poorer. There is also concern because the current policy is based on command and control and, once there is any relaxation of enforcement, it can easily be envisioned that those enterprises that were forced to shut down may secretly revive.[2] Therefore, it is necessary to consider how to set up a sustainable policy framework for emission reduction.

There are some studies on air pollution in China. Most of these studies have pointed out that China prefers the command and control style of dealing with pollution and that China has paid attention mostly to end-of-pipe measures in addressing air pollution. Many studies have suggested modification of this policy bias by implementing economic measures that provide incentives instead. However, few in-depth studies have been done on the background of the bias in Chinese policy.

Theoretically, economic measures have advantages in terms of economic efficiency. Given this, then, why has China, which is facing a severe shortage of capital investments for air pollution treatment because of the trade-offs that would involve with investments in economic growth, opted for a policy framework centred on command and control measures, which only present suboptimal efficiency compared to economic measures? Why has China opted to combine command and control measures and economic measures to form a policy mix? In the other studies, little effort has been made to answer such questions. The author believes the answer could be found by analysing the constraints on the enforcement of policy measures. This is the hypothesis that the author tries to prove in this chapter.

There are some studies on the enforcement of Chinese policies on air pollution (McElroy, Nielsen and Lydon, 1998; World Bank, 1997, 2001). These studies also analysed the background of Chinese policy bias based upon Chinese reality, and they pointed out that the source of the policy bias lies in the defects of the existing legal framework (i.e., although there is already a comprehensive legal framework, it lacks detail at the operational level and conflicts exist among different regulations), as well as the limited enforcement capacity (i.e., particularly in local governments, which serve as the main force in implementing air pollution policies) and the lasting influence of the planned economic system (i.e., the market economy, which is a precondition for the economic measures to be effective, is still immature). However, these elements are not sufficient to explain the causes of the policy bias. As indicated in the beginning of this paper, in reality command and control adopted by the Chinese government has contributed to reducing air pollution to a certain extent in the past few years. If the bias was caused by the factors that have been revealed by other studies, would not the command and control also fail to be implemented in the first place?

While considering environmental policies for developing countries, it is necessary to recognize the importance of the enforcement aspect, such as monitoring and administrative capacities (Fujikura in this volume; Matsuoka, 2000). Some studies have indicated that the high enforcement costs, including monitoring costs, have contributed to the policy bias in China. However, most of the studies failed to attach sufficient importance to such factors. In light of the energy consumption structure in China, more specifically, considering the coal distribution structure in China, the author believes great importance must be attached to finding measures to reduce the monitoring costs when proposing air pollution solutions in China. In this chapter, the author will put forward new policy options based upon this unique perspective.

This chapter covers the following contents. In Section 1, the author will analyse the cost effectiveness of two command and control measures, the 'Two

Control Zones' policy and the compulsory installation of FGD (flue gas desulphurization). The result shows such command and control suffers from very poor cost effectiveness. Next, in Section 2, the author will study the enforcement of an economic measure, the emission fee system. This study shows that, under the current conditions, such economic measures lack feasibility in terms of policy enforcement. In Section 3, based upon the analysis made in Sections 1 and 2, the author will further study why China has adopted command and control as its primary policy. Also, while providing alternative policy options, which the author believes are better than the current environmental policies, the author will explore the preconditions for such policy options to be effective. The alternative policy options include sulphur tax, an emission credit trade system and the setting up of public coal trading companies. In the final section, to conclude, the author will explain the difference in the policy options in comparison with other studies and the significance of this study. Also, while trying to generalize this empirical study, the author synthesizes the implications for the issues involved in effective environmental policy in a situation of imperfect market and government capacity, as is often seen in developing countries.

1. Current direct regulation and associated issues

Enforcement of the current command and control measures

'Two Control Zones' policy

China's primary air pollution prevention measure, the 'Two Control Zones' policy, was put forward in 1998 and became effective around 2002. In accordance with this policy, 175 cities (including cities on the county level and prefectural level) in 27 provinces were listed as key areas suffering from acid rain and SO_2 emission (Figure 5.1). The idea was to carry out effective pollution control by concentrating policy resources in those areas. Compared to the national target of reducing total SO_2 emission in 2005 by 10 per cent based upon the year 2000 level, the target for the 'Two Control Zones' was to reduce SO_2 emission by 20 per cent in the same period. Although the combined size and population of the Two Control Zones only amounted to 11 per cent and 39 per cent, respectively, of the national total, the zones' GDP accounts for 67 per cent and the SO_2 emission accounts for nearly 60 per cent of the national totals. Apparently, this measure aimed to produce results with significant cost effectiveness.

The following are the main aspects of the 'Two Control Zones' policy, although there are certain regional differences:

1. Within the target area, small coal-fired boilers are prohibited. Coal is not allowed in restaurants and transition to clean forms of energy (natural gas, LPG) is encouraged.
2. 'Zero Coal Areas' are to be established within the target area.

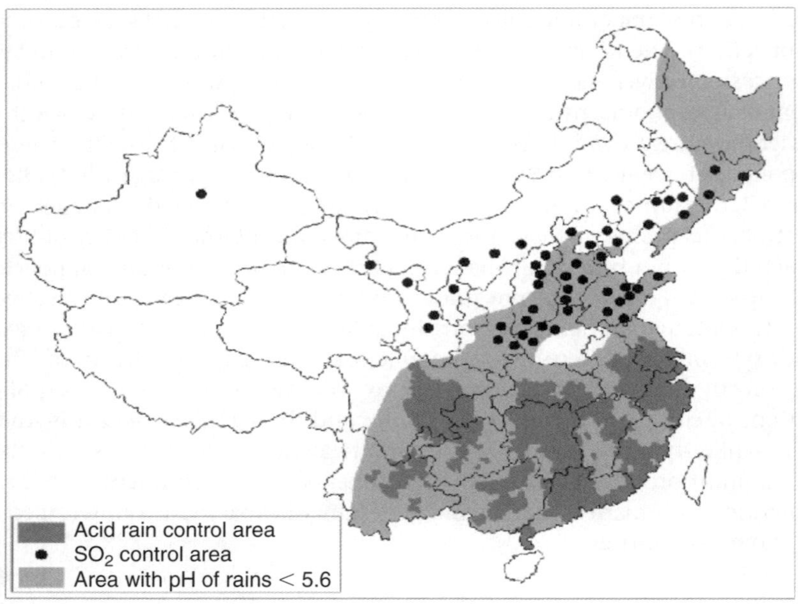

Figure 5.1   Distribution of the 'Two Control Zones'
Source: IEA, 2002.

3. High-sulphur coal is prohibited by restricting the sulphur content of coal
   within the target area.
4. All new power plants are requested to install FGD.
5. All new emission units within urban areas are requested to install online
   Continuous Emission Monitors (CEMs).

All the above-mentioned measures are command and control measures.
Usage of coal-by small polluters is completely prohibited, and mandatory coal
quality standards are imposed on medium and large polluters. The polluters are
not able to choose the most cost-effective approach to address pollution. The
practice wherein policy-makers designate the specific measures is referred to as
'setting technical requirements', which differs from pollution control by setting
up emission standards and leaving it to enterprises to find ways to meet such
standards, although both fall within the range of command and control. In reality, cost effectiveness has been impaired since policy-makers, who are not well
informed of such relevant information as pollution abatement costs, make
decisions on the specific measures on behalf of the polluters.

We will further discuss the above-mentioned points 2 and 3 under the
next heading and point 4 under the following one.

Quality standards for sulphur content of coal

There are some defects in the 'Two Control Zones' policy from the perspective of cost effectiveness. There are different options for reducing SO$_2$ emission, including improving coal quality (and, of course, other forms of fuel, such as natural gas, could also be considered), adopting CFBC (circulating fluidized bed combustion) technology, implementing end-of-pipe measures, like installing FGD, or using cleaner production approaches. To identify the most cost-effective approach it is necessary to consider such variables as coal consumption volume, quality, price, investment needed for new technology and the operational costs. In addition, risk assessment and consideration of financing possibilities might also be necessary for the FGD equipment that requires massive investment. Obviously, conditions vary greatly among the various enterprises. However, current Chinese policy imposes the same approach despite the complex reality: prohibiting coal in small enterprises, setting up coal quality standards for mid-size enterprises and requesting large enterprises to install FGD.

The primary feature of this policy is to drive coal, particularly high-sulphur coal, out of the control zones. However, such a practice might exert a negative impact on cost effectiveness. Because the policy inhibits demand for coal with high-sulphur content, some coal mines producing low-quality coal in the neighbourhood might be forced to stop production. In reality, considering the transportation costs, it could be a better solution for some polluters within the control zones, to install end-of-pipe measures and continue to consume local high-sulphur coal. To offer another example, by installing FGD, the enterprises could sell a certain volume of gypsum as a by-product; hence it is most profitable to consume coal with a sulphur content of 2–3 per cent. However, such cost effectiveness cannot be realized since high-sulphur coal is prohibited within the control zones and production of high-sulphur coal is therefore suppressed.

This policy framework provides no incentive for enterprises to develop technologies to reduce emission at lower costs. Policy-makers should also consider the dynamic effect on cost effectiveness, not only the static effect. Technical innovations can reduce abatement cost and improve cost effectiveness. The current 'Two Control Zones' policy framework does not provide incentives to make such technical innovations happen, as polluters only need to meet the requirements on coal standards. Because the above-mentioned technical standards aim at controlling the coal quality, instead of emission volume, only coal-washing technologies are affected by this policy framework, and it is not conducive to the development of new technologies that could fundamentally reduce emissions.

Policies for enforcing installation of FGD

The Chinese government attaches great importance to installation of FGD in power plants as one of the key measures for pollution control in the future. This policy is not confined to the 'Two Control Zones'. In accordance with the decree issued on 30 January 2002, power plants need to adopt the following measures.

Power plants that fall within 1 through 3 need to install FGD: (1) power plants fuelled by high-sulphur coal; (2) new power plants and power plants intending to expand capacity; (3) existing power plants that cannot meet the SO_2 emission standard or total emission standard and whose remaining lifetime exceeds ten years. Power plants that fall under (4) must consume low-sulphur coal or adopt measures that make it possible to achieve the same emission reduction target, which is to install a simplified FGD: (4) existing power plants that cannot meet the SO_2 emission standard or the total emission standard and whose remaining lifetime is less than ten years. Although the regulation does not cover those power plants that meet the emission standard and those power plants are merely requested to adopt the above-mentioned environmental protection measures, in fact, many power plants fall within the range of 1 through 4, and these are all forced to install FGD.

Furthermore, (1) power plants are requested to consume coal with a sulphur content of less than 2 per cent, the units with installed capacity exceeding 200 MW must adopt wet-type limestone-gypsum as a priority, the desulphurization rate needs to exceed 90 per cent and the operation hours of FGD must exceed 95 per cent of the operation hours of the power generating equipment; and (2) small and medium power plants (installed capacity less than 200 MW) fuelled by coal with not more than 2 per cent sulphur content or a remaining lifetime of less than ten years must adopt low-cost semi-dry-or dry-type mature technologies, achieve a desulphurization rate exceeding 75 per cent, and must ensure that the operation hours of FGD exceed 95 per cent of the operation hours of the power generating equipment.

On the other hand, the following measures were put forward in order to control emissions from industrial and residential boilers: (1) small and medium boilers less than 14 MW are encouraged to consume briquette coal, low-sulphur coal and washed coal, and install a wet-type dust catcher and a simplified FDG; and (2) medium and large boilers more than 14 MW are requested to consume low-sulphur coal instead of high-sulphur coal, adopt CFBC technologies and install an FGD. However, these measures were not enforced until 2003. In addition, small boilers of less than 4 tons in urban areas were forced to suspend operations and were dismantled. It is projected that the Chinese government plans to phase out small boilers that cause severe pollution and gradually enforce the above-mentioned regulation in order to reduce pollution caused by the other remaining boilers.

As to the cost effectiveness of the above-mentioned measures, it is difficult to judge whether optimal emissions reduction could be achieved since such measures are similar to regulation of coal quality in the form of technical requirements. Considering the fact that power plants have been forced to install apparently high-cost equipment from the end-of-pipe desulphurization equipment menu, it could be presumed that cost effectiveness is far from optimal. Below, the author has made a rough calculation of the total investment needed if all coal-fired power plants that existed in 2000 were to install FGD (Table 5.1).

Table 5.1   Cost calculation on installing FGD in all coal-fired power plants

	No. of units	Installed capacity (MW)	Technology	Initial cost (1,000 yuan)	Annual running cost (1,000 yuan)
Units ⩾200 MW	452	1,259.25	Wet-type limetone-gypsum	84,243,830	26,190,240
Units <200 MW	2,467	725.40	Dry-type LSD method	34,529,040	41,946,400
Total	2,919	1,984.65		118,772,870	68,136,640

Note: The data for number of units and installed capacity are from 2000 (*Yearbook of Power Industry of China*, 2001 edn). The initial cost, annual running cost, wet-type limestone-gypsum FGD were calculated as unit investment of 669 yuan/kW and 57,943,000 yuan/set. Dry-type LSD equipment was calculated as unit investment of 476 yuan/kW and 17,003,000 yuan/set (the above data were based upon 1995 prices, compiled by Wang *et al.*, 2002, p. 78).
Source: The author made the calculation based upon the above-described data. In addition, based upon the deployment rate of FGD in 1998, those already installed among power generation units above 200 MW were deducted from the calculation.

The table shows that the initial cost would be as high as 118.8 billion yuan if all power plants were to install FGD, which is equal to 1.3 per cent of the GDP in 2000 and 3.6 per cent of the national fixed asset investment for the year. It also shows that the running cost would reach 68.1 billion yuan, which means that each year huge additional amounts of capital would need to be invested. In the future, the construction of power plants in China will be accelerated, and according to one projection, the installed capacity of both coal-fired and gas-fired power plants in China will reach 302.35 GW by 2010 and 479.61 GW by 2020, 1.4 times and 2.3 times higher, respectively, than that of 2000 (Zhou *et al.*, 2003). Considering that such massive investment is needed in power generation equipment, it is doubtful whether the investment in FGD equipment can keep pace. Therefore, the policy should allow enterprises to adopt measures other than FGD in accordance with the principle of cost to performance.

Regarding the dissemination of FGD, statistics show that, across the world, 226.82 GW worth of installed power generation units have installed FGD in total (1998 statistics, compiled by Wang *et al.*, 2002, p. 55). Since the global installed capacity of thermal power plants equals 2074.47 GW (including oil and natural gas-fired units), this means that only 10.9 per cent have installed FGD. Even in OECD countries, only 20.1 per cent of the combined installed capacity of 1126.98 GW has installed FGD. In fact, Japan and United States are the only two countries in the world where FGD has been introduced on a considerable scale. Thus, it is a genuine cause for concern that the installation of FGD by regulation in all power plants in China will greatly increase the emission reduction cost and impair cost effectiveness.

From the perspective of impact on achieving dynamic cost effectiveness through technical innovation, the above-mentioned policy also causes certain concern. Since FGD is the only option imposed by this policy, the other emission reduction technologies, such as cleaner production measures, might be negatively affected. Although the policy might promote reduction of the production cost of FGD and, thus, improve the cost effectiveness of this policy, most of the R&D resources in the society may be pulled to concentrate on FGD technologies. This would negatively impact R&D on other technologies with better cost effectiveness. From this perspective, in terms of providing incentive to technical innovation, this policy framework entails a considerable drawback because it enshrines FGD as the only emission reduction approach.

The major concern about the command and control policy is its cost effectiveness. Below, the author will illustrate the cost-effectiveness problems associated with such command and control by reviewing pollution abatement projects implemented in the 'Two Control Zones'.

Cost effectiveness of direct regulation

Data used for the analysis are from the investment amount and projected SO_2 reduction shown on the List of Air Pollution Abatement Projects to be implemented in the tenth five-year period within the 'Two Control Zones'. The investment amount and SO_2 reduction volume are derived from rough initial calculations, but considering the fact that there have not been any comparisons of costs and performance covering all projects in the country, the analysis conducted here is rather significant. In addition, the projects on the list were all to be implemented. In that sense, the list is a genuine reflection of the policy preferences of the Chinese government in the area of environmental protection.

There was a total of 279 projects on the list, and the projects are classified according to their technology approaches, as shown in Table 5.2. We can see that projects for FGD top the list, followed by urban natural gas projects, and this shows the impact of the two pillars of the guideline policy or the 'Two Control Zones', which stipulates natural gas for residential purposes and FGD for power plants. It is worth noting that few projects were related to coal washing. In Figure 5.2, the author graphs all the projects in terms of their investment amount and SO_2 emission reduction and, as shown on the graph, most of the larger projects are FGD projects and some are urban natural gas projects and CFBC projects.

The correlation between investment amount and emission reduction volume was calculated to be 0.53. Considering the inherent relationship between the two factors, the resulting figure seems to reflect a lack of proper co-relation. This figure implies that investment for abating air pollution failed to bring about the expected emission reduction. The calculation seems to support the conclusion drawn by the author, which is that the command and control measures have significant drawbacks in terms of cost effectiveness.

Table 5.2   Projects under the 'Two Control Zones' policy framework

	Options	No. of projects	Investment
1	Coal washing	6	3.5
2	Simplified FGDs	50	18.5
3	FGDs	80	257.6
4	FGDs into small boilers	25	22.0
5	Shutdown of small boilers	1	0.6
6	LPG	11	9.1
7	Urban natural gas projects	23	42.2
8	Fuel switching	6	4.8
9	Coal quality improvement	3	0.2
10	CFBC	12	10.2
11	Coal gas	8	7.0
12	Others	54	35.1
Total		279	410.8

Source: Compiled by the author, based upon the 'List of Key Abatement Projects for 10th Five-year Plan Period in the Two Control Zones'.

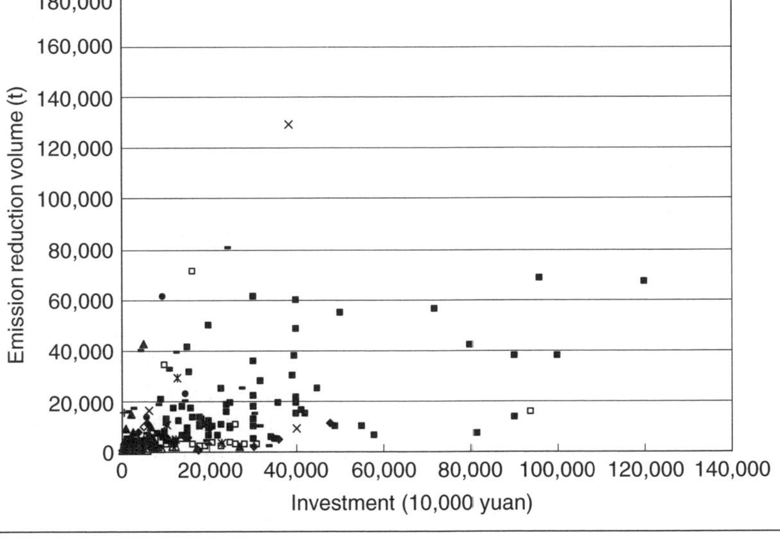

Figure 5.2   Distribution of investment and SO$_2$ reduction
Source: Compiled by the author, based upon the 'List of Key Pollution Abatement Projects in 10th Five-year Plan Period in the Two Control Zones'.

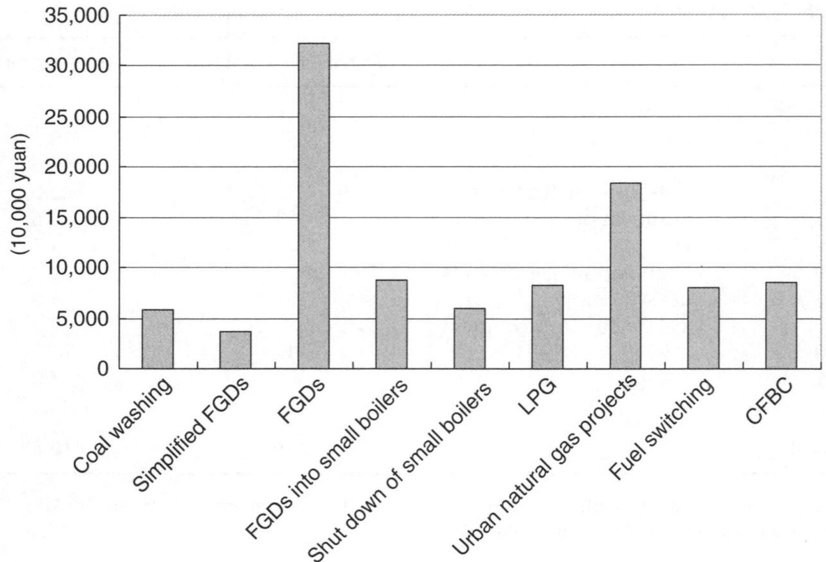

Figure 5.3   Average investment per project for various pollution abatement options
Source: Compiled by the author, based upon the 'List of Key Pollution Abatement Projects in 10th Five-year Plan Period in the Two Control Zones'.

Figure 5.3 shows the average investment amounts for various options, and it clearly shows that the average investment amounts needed for FGD and urban natural gas projects by far exceed the investment needed for other options. However, if one analyses the cost of each option and the performance-unit emission reduction cost of SO_2/ton, as shown in Figure 5.4, then the efficiency of FGD and the urban natural gas projects is not necessarily higher than the other options. The inefficiency of the improvement of coal-fired boilers (4 and 5) could be explained by the lack of economy of scale. As to the poor cost efficiency of urban natural gas and LPG projects, this could be explained by the fact that, on the one hand, these projects require a huge investment to build the pipeline network and, on the other hand, residential consumers, who are the main users, do not emit so much SO_2. It is worth noting that coal washing, simplified FGD and CFBC projects achieve excellent cost efficiency.[3] The issue here is why the emphasis was on FGD and urban natural gas projects instead of coal washing and CFBC projects, which enjoy better cost effectiveness.

Figure 5.5 shows the SO_2 emission reduction cost and SO_2 emission reduction volume for coal washing, FGD and urban natural gas projects, respectively. Since in decision-making, in addition to cost effectiveness, the achievable reduction volume associated with the option is another important factor, it is reasonable to evaluate the policy framework by examining the balance between

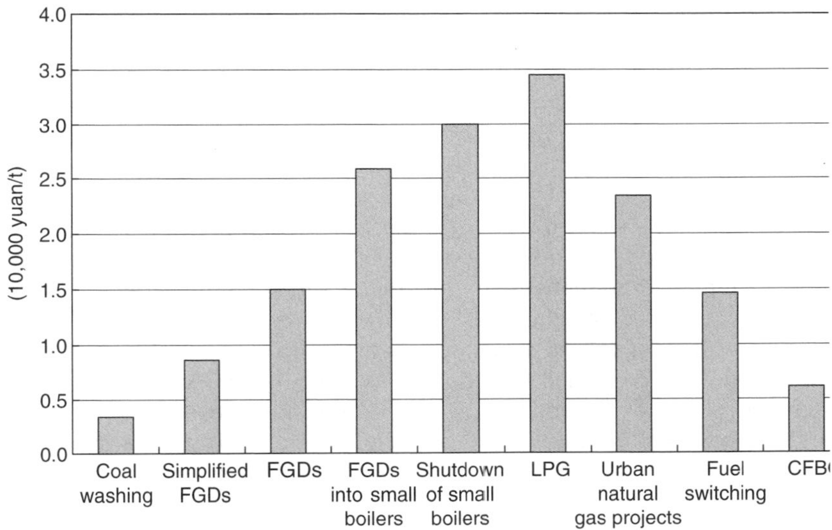

Figure 5.4   Unit cost of various pollution abatement options (weighted average)
Source: Compiled by the author, based upon the 'List of Key Pollution Abatement Projects in 10th Five-Year Plan Period in the Two Control Zones'.

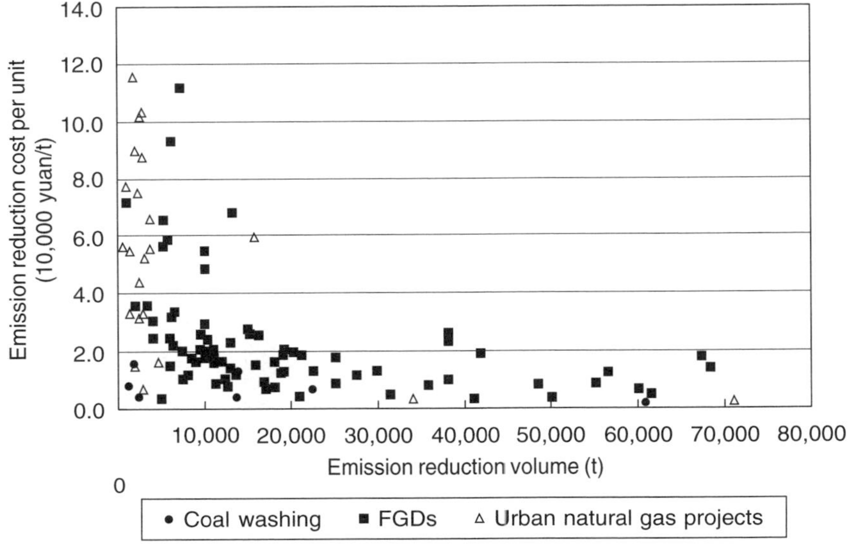

Figure 5.5   Cost and emission reduction volume of three options
Source: Compiled by the author, based upon the 'List of Key Pollution Abatement Projects in 10th Five-year Plan Period in the Two Control Zones'.

the horizontal and vertical axes. The figure illustrates that the reduction volume associated with FGD (horizontal axis) is widely distributed. On the other hand, in terms of cost to performance (the vertical axis) emission reduction costs concentrate at around the level of 30,000 yuan/ton; however, there are still projects with very poor cost effectiveness. As for urban natural gas projects, only limited emission reduction could be expected, and the emission reduction cost of a great number of the projects exceeds 30,000 yuan/ton. In contrast, there were a mere six coal washing projects, although the emission reduction cost of such projects was only around 10,000 yuan/ton and the expected reduction volume was also higher than the urban natural gas projects.

To a large degree, the effects of the FGD projects are widely spread, both in terms of reduction volume and cost to performance, because the power generation scale and coal quality consumed by power plants varies. This is fully reflected in Figure 5.6, which shows the relationship between various options and the cost performance in different regions. Southern China, where low-quality coal was consumed, has shown better cost effectiveness. Compared to Northern China, the cost effectiveness of coal washing, FGD and urban natural gas projects in southern China is significantly higher.

The following conclusions could be drawn from the above simple analysis. First, under command and control measures, options chosen for pollution abatement do not produce cost-effective results. Even when the same options were applied, significant cost-effectiveness gaps are observed due to varying

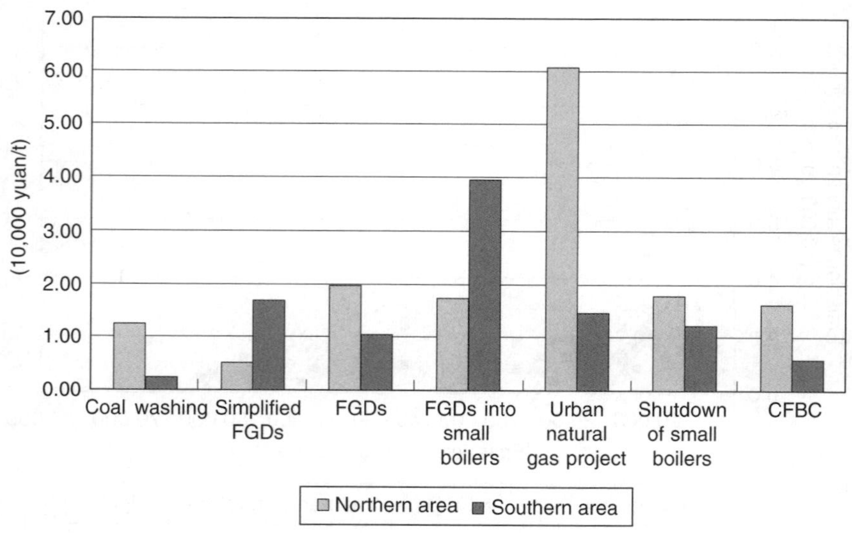

Figure 5.6   Unit emission reduction cost of various options in different regions
Source: Compiled by the author, based upon the 'List of Key Pollution Abatement Projects in 10th Five-year Plan Period in the Two Control Zones'.

local conditions. Urban natural gas projects, in particular, are far from an optimal option, both in terms of cost effectiveness and reduction volume achievable.[4] However, in reality, a great number of such projects are being actively pursued.

While discussing why the policy framework failed to reflect cost effectiveness, it is worthwhile to understand how the above-mentioned project list was produced. Since the 'Two Control Zones' policy aims to control the total emission volume and sets a target of 20 per cent reduction in SO_2 emission by 2005 based upon the emission level of 2000, naturally it affected the way the project list was made up. It is assumed that, after establishing the reduction targets based upon the total emission volume control policy, each province made its list using a top-down approach to meet the requirements. During this process, primary attention was paid to accumulated reduction volume and little attention was paid to the cost effectiveness of the projects. The list clearly reflects how the projects were scrambled together behind the scenes with little rationalization. Since the general policy guideline is to shift from coal to natural gas in urban areas and to adopt FGD measures, even though these two options are not optimal in terms of cost effectiveness, understandably they would be the favoured choices for the list.

It would be hard to avoid biases against cost effectiveness under such a decision-making process. The specific projects were determined in a top-down manner to satisfy the target of 'Two Control Zones' policy. Although local governments are closer to polluters, in fact they have little information on the emission reduction costs of each emission source, and this information is necessary to enable them to choose the optimal projects for the list. The compulsory target of 20 per cent emission reduction in the 'Two Control Zones' itself neglected the fact that each area has different abatement cost curves. This shows that this policy paid little attention to cost effectiveness.

The above analysis shows that the current command and control measures may result in poor cost effectiveness and high costs for emission reduction. Then why did the Chinese government adopt this policy? The author further suggests that it has something to do with the constraints of policy enforcement in present-day China, as shown in the following section on enforcement of the current economic measure, the emission fee system.

2.   Current economic measures and associated issues

Of all the environmental policies adopted in China, the emission fee, which is a kind of surcharge on the emission volume of coal dust and SO_2, is one of the few that is an economic measure. The following issues have been pointed out in association with the emission fee policy:

1. Since the surcharge rate is much lower than the marginal abatement cost, polluters are more inclined to pay the surcharge than to apply pollution

abatement measures, and this exerts negative incentives on the promotion of emission reduction measures.

2. The emission fee is imposed only on the part of emission volume that exceeds the emission standard, not on the total volume of discharged pollutants.
3. Of all the pollutants subject to the emission charge, polluters only need to pay for the predominant type of pollutant, and no fee is collected for other types of pollutants.
4. Contrary to the initial regulation (State Administration of Environment Protection, 2003b), part of the revenue from the emission fee is allocated for projects not associated with environmental protection purposes.

In light of the above issues, it is doubtful whether the emission fee policy implemented so far has fully played its role as an economic measure. In July 2003, significant reforms were carried out in relation to the emission fee system, including:

1. As of 2004, the SO_2 emission fee shall be increased from the current 0.2 yuan/0.95 kg to 0.4 yuan/0.95 kg and, as of 2005, it will be further increased to 0.6 yuan/0.95 kg.
2. The total emission volume shall be subject to the emission fee.
3. Polluters have to pay emission fees on all types of pollutants.
4. All collected emission fees must be handed over to the fiscal departments for strict management and all applications of the funds are subject to examination and approval.

Despite the above-mentioned reforms, there are still defects associated with the enforcement of the emission fee policy. We will discuss the real enforcement of the emission fee system in the following section.[5]

The emission fee should be collected based upon accurate monitoring of the SO_2 emission. However, the Environmental Protection Bureaus on the county level that are supposed to enforce this policy are unwilling to invest in the necessary monitoring facilities using their own budget because of the lack of additional financial support for investment in monitoring facilities. Therefore, they only require enterprises to report the ash content and sulphur content of coal they consume, and then they calculate the coal dust and SO_2 emissions based on those reports to determine the amount of the charge. Since false reporting cannot be completely eliminated, it is necessary to carry out inspections, including sampling. However, since such inspection expenses also need to come from the local Environment Protection Bureaus' own budget, such inspections are rarely carried out. Because of this, there is a certain amount of room for negotiation between the authorities and enterprises when determining the amount of the fee. On the other hand, it is by no means an easy task to collect the emission fee. Some enterprises with low profits even resist with force attempts to collect the fee. As a result, the emission fees tend to be reduced, and the authorities can only collect the amounts that are affordable to the enterprises. A considerable number of enterprises pay nothing at all. According to one survey, the emission fee collection rate was 70 per cent above the county level and 50 per cent below the county level. The authorities failed to collect half of the emission fees below the county level (Li, 1999, p. 108).

The emission fee collected is supposed to be allocated as follows: approximately 70 per cent is refunded to the enterprises if they invest in the pollution abatement projects, 10 per cent goes to the national treasury and the remaining 20 per cent is apportioned to the budgets of the local Environment Protection Bureaus on the provincial and county level. Here, there is a problem. Particularly below the county level, budgets of local Environmental Protection Bureaus come, in large part, from the emission fee, and for them, the emission fee is a fixed source of revenue. They simply do not want this revenue to be reduced due to the pollution abatement projects implemented by the enterprises. According to surveys of coal users, it is often mentioned that the emission fees are collected based upon the amount of the previous year and, for example, there was a case in which the amount of the fee was not reduced even after an enterprise shifted to low-sulphur coal.

According to the regulation, the collected emission fee should be refunded if an enterprise applies for funds for an environmental protection project. However, the range of environmental protection projects entitled to refunds is so wide that many cases have been heard of where refunds were obtained for landscaping or for improving employee welfare facilities in the name of environmental protection projects. Therefore, we cannot fully expect this policy framework to play its role in promoting pollution abatement projects. Furthermore, in principle, the collected emission fee is only refundable to the enterprise that paid it and may not be refunded for other projects addressing pollution sources. It is often pointed out that such investments become tax-deductible if they are appropriated as production costs, and in fact the traditional emission fee is viewed as funds accumulated for future equipment investment (Blackman and Harrington, 1999).

In summary, looking at the actual enforcement of the emission fee policy, the stated objective of this surcharge system, which aimed at encouraging enterprises to implement pollution abatement measures by linking the surcharge to emissions, exists in name only. The policy has evolved into a sort of fixed tax that is only tenuously related to the emission volume of coal dust and SO_2. It is not working as an incentive to encourage enterprises to actively pursue emission reduction measures. This situation is a result of the limited capacity of the Environment Protection Bureaus on the county level in terms of monitoring the actual emission of the polluters. The result shows that the reality is far removed from the predicted theoretical outcome, and it is due to the constraints of enforcement costs that policies such as this one fail to achieve the expected result. Instead of providing incentives for enterprises to reduce emissions, the primary role played by the emission fee policy prior to 2003 was as a fund for enterprises for projects not directly related to environmental improvement and as a funding channel for local Environment Protection Bureaus.

Generally, it is believed that when comparing command and control measures with economic measures, including the emission fee, the economic measures are better at promoting technical innovation. In fact, as an institutional framework, the surcharge system should be able to promote technical innovation and achieve dynamic benefits. However, because the threshold for the emission fee was very low before the reform, many enterprises believed that so long as they paid the fee, there was no need to take further measures. Thus, the system has conversely hampered technical innovation. Understandably, the problem did not originate from the system itself, but from the way the fee rate was set. When applying the surcharge in reality, it is impossible to set the rate so that it equals the optimal cost effectiveness in theory. From this perspective, it should be realized that when it comes to actual introduction of the policy, one cannot ensure the promotion of technical innovation through improvement of dynamic efficiency, although this is believed to be a leading benefit of economic measures. This is particularly true in those developing countries that are suffering from lack of policy enforcement capacity and insufficient enterprise information disclosure.

The phenomenon of violation of fairness, like refusal of payment by politically influential enterprises, can be seen as a result of the limited administrative capability of the Environment Protection Bureaus on the county level. Furthermore, since they are not capable of monitoring the actual emission volume, they cannot capably administrate and therefore fail to collect the emission fees from a great number of polluters. In addition, in the local political structure, the power of the Environmental Protection Bureaus is rather weak, and they depend on the local government, which consists of local politicians. Often the local politicians are more concerned about developing the local economy and protecting the local enterprises than about environmental protection. Under such circumstances, the Environmental Protection Bureaus neither have the will nor the power to oppose the wills of local politicians in enforcing the policy.

The new emission fee system implemented in 2003 is a great step forward in the right direction. In particular, the rise in the threshold and the shift to charges based on total emissions could make the economic incentives work. In addition, collected fees are not only available as refunds to the enterprises that paid them but can also be used for other environmental protection projects. This could accelerate the investment in environmental measures. However, without emission monitoring and sufficient enforcement capacity, the reformed policy cannot be smoothly implemented. Based upon our analysis, there are considerable challenges in this area. Consequently, the new policy may have a certain impact on large enterprises that are easier for government to monitor, but when it comes to small and medium polluters, there are still considerable difficulties in implementing the policy. In China, SMEs cause a huge amount of pollution, and they should by no means be neglected. Furthermore, there is the potential risk that the implementation of the reform might further impair the emission monitoring and enforcement capacities of the local Environment Protection Bureaus.[6] Continuous attention should be paid to the future of the reformed policy.

3. Policy planning to mitigate air pollution in China

Background of bias towards command and control

The biggest cause of air pollution in China could be attributed to its energy structure, in which coal occupies a dominant share. Coal contributed to a great proportion of the total emission of pollutants, including 87 per cent of SO_2, 60 per cent of dust, 67 per cent of NOX and 85 per cent of CO_2, respectively (Chen, 2001). Therefore, to address air pollution in China, it is critical to deploy clean energies, including natural gas. However, it is unrealistic and economically unacceptable to abandon coal simply because it causes great air pollution. Furthermore, as described in this chapter, the viability of a fuel switch away from coal is impaired by high economic cost and poor cost effectiveness. It is predicted that, in the future, coal will remain the primary source of energy in China.

Table 5.3 compares the price difference of raw coal and clean energies for end users in Hunan Province. The table shows that to obtain the same heat value level, despite improvements in combustion efficiency, the price of natural gas is 2.5 times more expensive than that of raw coal. Of course, for those who can afford it, such as urban residents, shifting to natural gas is a feasible option. However, for many industrial enterprises, a price hike of 2.5 times in fuel cost is simply unacceptable. Therefore, it is important to adjust measures in line with reality to optimize the way coal is consumed. For the same coal, if clean

Table 5.3   Price gap between raw coal and clean energies in Hunan Province (2003)

	Price (yuan/t)	Heat content (MJ/kg)	Combustion efficiency (per cent)	Actual combustion heat content (MJ/kg)	Price of same heat content (yuan/GJ)	Price ratio
Raw coal	262	22.13	60	13.27	19.74	1
Washed coal	340	23.00	60	13.80	24.64	1.25
CWM	430	18.82	95	17.89	24.03	1.22
Heavy oil	2300	40.17	98	39.37	58.42	2.96
Natural gas	1.8*	35.98	98	35.26	49.93	2.53

Note: * Yuan/m^3.
Source: Compiled by the author, based upon field survey.

coal technologies (CCT) such as coal washing and coal water mixture (CWM) could be introduced, the price increase would be only around 20 per cent, while at the same time the emission level could be reduced to meet the environmental standard. Depending upon the specific situation of the users, there are additional options, including CFBC and FGD, etc. However, in reality, coal washing and CWM are only used in limited areas due to price gap between the processed coal and raw coal.

Ultimately, there are two factors that make it so difficult to disseminate CCTs in China: on the one hand, the environmental protection policies have not been effectively implemented; on the other hand, raw coal is available at an extremely low price. If the environmental regulations had been very effective, the consumption of raw coal would be restricted, as it exerts massive pressure on the environment. The reality, as analysed in Section 2 on the implementation of emission fee system, is that the emission fee regulation failed to deliver the desired effects and the consumption of raw coal has been affected very little. Therefore, it is imperative to enhance the effectiveness of the environmental regulations and raise the price of raw coal to reflect the pollution it causes to the environment.

While considering specific ways to boost the effectiveness of environmental regulations, lessons can be drawn by analysing the implementation of the emission fee policy. In fact, the lessons that can be drawn are also related to China's decision to adopt command and control measures, particularly the rigid technical requirements, as its primary type of environmental protection policy. Even if the authorities wanted to adopt economic measures similar to the emission fee or to regulate emission by way of emission standards, the prerequisite is to be able to monitor the emissions from enterprises. Such monitoring is almost impossible for the local Environmental Protection Bureaus on the county level, who are assigned to implement the policy. On the other hand, if technical requirements are adopted, the authorities only need to check whether the equipment has been installed and is running; little further monitoring is required. The mandatory policy that all power plants install FGD shows that the government fully realizes the importance of addressing emissions from power plants, which account for 42.6 per cent of the SO_2 emissions by the industrial sector. However, the cost of such a policy is much too high from the perspective of the cost to performance principle, and the calculations in Table 5.1 support this conclusion. It is assumed that one of the important factors that led Chinese authorities to adopt this policy is that the enforcement of technical requirements is much easier than monitoring.

On the other hand, it was also out of consideration for policy enforcement effectiveness that the quality standard for the sulphur contents of coal, instead of an emission standard for emission volume, was adopted in the end. Of course even with regulations on coal quality, inspections such as sampling testing are necessary. Since they are also costly, it is assumed that in many cases the local Environment Protection Bureaus do not carry out such inspections. In that sense, the monitoring is not adequate. However, the monitoring of coal quality costs less than monitoring of emission volume. In fact, it is assumed that the 'Two Control Zones' policy was adopted also in light of significant constraints on local enforcement authorities, namely the Environment Protection Bureaus on the local level, in terms of personnel and funding. By concentrating policy resources only in the control zones, the effectiveness of the policy can be improved and, in this sense, the 'Two Control Zones' policy is more pragmatic and has certain merits.

However, there is a concern that since the 'Two Control Zones' policy tries to suppress the circulation of high-sulphur coal within the zones by regulating the coal quality, it might have a negative impact by causing some to try to avoid the regulation. The cities listed in the 'Two Control Zones' are mostly urban areas of medium and large size, and enterprises within the zones have already implemented environmental protection policies that exceed those of enterprises elsewhere. It is very likely that there will be an outflow of high-sulphur coal to enterprises outside the 'Two Control Zones' since only low-sulphur coal is allowed within the zones. Generally speaking, enterprises located on the outskirts of the city are rather small and have only poor environmental protection measures. Therefore, the result might be that pollution is transferred to the areas outside the control zones, and the total emission volume of the society overall might be greatly expanded.[7]

In summary, based upon the judgement that it is more important to increase the effectiveness of the policy enforcement than the cost effectiveness, the Chinese government has established a policy framework to promote emission reduction that primarily depends on command and control measures consisting of technical requirements and supported by economic measures such as the emission fee system. It is rather doubtful whether such a policy framework is sustainable in the future because it neglects cost effectiveness, many polluters will try to evade such a policy and it might cause a heavy impact on the economy. Therefore, it is necessary to introduce policy measures that will reduce the implementation cost, provide sufficient flexibility to the enterprises and enhance cost effectiveness.

In the following section, the author will make a few suggestions for more desirable alternative options to the current environmental policies in China. The alternative options are sulphur tax, an emission credit trade system and setting up of public coal trading companies. While discussing the specific policy options, the conditions for making measures for air pollution control in China work will be revealed.

Alternative policy 1: sulphur tax

According to certain theories, for those developing countries constrained by inadequate monitoring and enforcement capacities, the introduction of an environment tax is recommended (Matsuoka, 2000). Generally, governments refrain from fuel tax because it is regressive, or the poorer the user, the greater the burden becomes since everyone needs fuel in daily life. In most cases, the tax is imposed on end users, with small end users, including residential end users, being excluded (OECD, 2002). However, for China, in light of its coal distribution structure, it would be more desirable to impose sulphur tax according to the sulphur content of the coal at the time coal is shipped from coal mines.

In China, around 28,000 coal mines (2003) provide coal as fuel to 4,300 coal-fired power generation units (2003) and over 500,000 industrial and residential boilers (1998) (which alone account for 75 per cent of overall coal consumption in China). If tax is imposed on coal mines instead of on end users, the number of monitoring points will be reduced to 1/25. As for sampling of coal quality, compared to inspections on each end user, it would greatly reduce inspection costs if inspections were carried out in the coal mines. Furthermore, ultimately the sulphur tax is borne by enterprises as end users, and so the coal mines just need to add tax to their normal prices and then the administrative authorities would not have to deal directly with the end-user enterprises. This would help to prevent tax evasions.

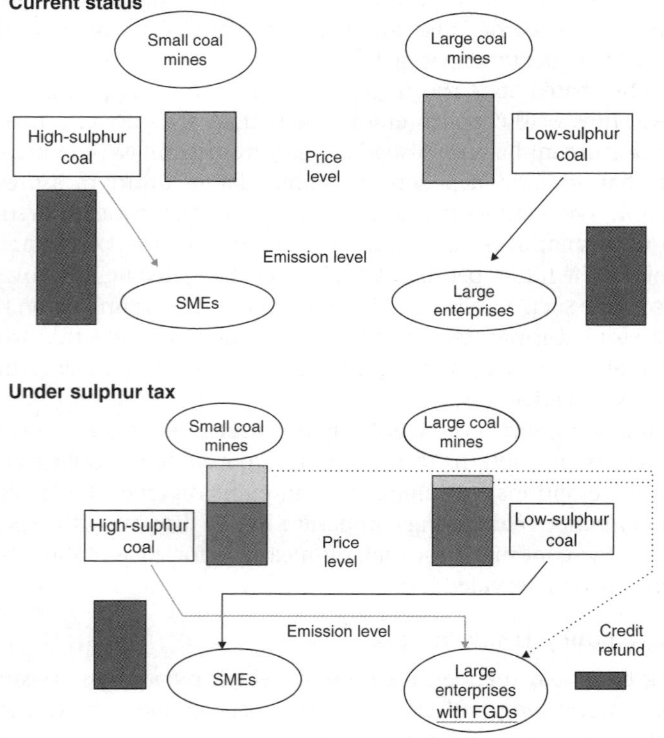

Figure 5.7   Impact of sulphur tax

Figure 5.7 illustrates the impact of sulphur tax. In the current Chinese coal market, since the emission fee system has not been very effective for emission reduction, the sulphur content of coal has had little impact on coal prices and, instead, the prices have been determined more by the production and transportation costs. The coal distribution pattern features a dual structure of large state-owned coal mines supplying large enterprises and local small and medium coal mines supplying small and medium enterprises. Within the same region, coal supplied by small and medium coal mines is cheaper than the former. In terms of transportation, large state-owned coal mines distribute coal across the country by railway, and small and medium coal mines mainly target local markets within a very small region (Horii, 2003). Under such a market situation, coal consumption in China is characterized by small and medium enterprises that consume the cheap high-sulphur coal supplied by local small and medium coal mines and by large enterprises that consume the high-quality low-sulphur coal supplied by large coal mines located in northern China. This is particularly true in the southern part of China, where only low-quality coal is produced and acid rain and SO_2 emissions are extremely severe. As the result, emissions from SMEs, which have insufficient measures for pollution control, are very high, resulting in a higher national emission level.

If sulphur tax is introduced, the prices of high-sulphur coal produced by local small and medium coal mines will increase significantly, while the prices of low-sulphur coal produced by large coal mines will only increase marginally, and the relative prices of the two will change significantly (Figure 5.7). Consequently, prompted by cost considerations, SMEs will begin to purchase low-sulphur coal produced by large coal mines. Of course, if no other measures are taken, the sulphur tax will only result in an increased stock of high-sulphur coal in the market. This is not desirable from the viewpoint of the economic utilization of high-sulphur coal, and this also hampers the dissemination of CFBC, FGD and other desulphurization technologies. Therefore, it is recommended that a system be established that could refund part of the sulphur tax revenue as credits to enterprises that apply desulphurization technologies. Those enterprises planning to apply desulphurization technologies would have to file an application with environmental protection authorities for credits and, in compensation, those enterprises would be obligated to accept strict CEM monitoring. The amount of credits should be calculated based upon the consumption volume of high-sulphur coal by the enterprise in the past and the amount of desulphurization based upon the monitoring system. Using the credits, enterprises could purchase high-sulphur coal and, thus, high-sulphur coal would not flow out to small and medium enterprises.

The biggest benefit of the sulphur tax lies in the fact that it would virtually bypass the high enforcement costs associated with monitoring the SMEs and the collection of surcharges. With other forms of economic measures, for instance the emission credit trade system, it is necessary to monitor the actual emissions from enterprises. Even with command and control measures that use emission standards, it is still necessary to monitor the actual emission. Considering the number and the range of distribution of SMEs, the monitoring cost would be massive.

Certainly introduction of a sulphur tax would also require monitoring of the coal quality in coal mines through sampling; however, as described earlier, on the one hand the number of monitoring points would be drastically reduced and, on the other hand, concentrated sampling could be carried out. Therefore, monitoring could be carried out with significantly fewer costs. At present, few samplings and inspections are conducted since the local Environmental Protection Bureaus do not want to shoulder the expense of monitoring. In terms of tax collection, whereas collecting directly from the end users results in great amounts of taxes being uncollectible, sulphur tax would be collected at the coal mines; with no need to face the end users, the uncollectibles would also be significantly reduced.

Under such a system, the fairness of the marketplace could be improved because SMEs, which have shared little of the external environmental costs in the past, would share part of the costs. Of course, to refund the tax revenue in the form of credits also means to make allowances to the polluters, hence there will be a certain normative issue. However, such a credit refund also means that expensive investment in desulphurization technologies are shared by the whole society. The public should view the practice of refunding credits based upon the market mechanism favourably since in the future it is imperative to introduce end-of-pipe solutions, such as FGD. It has been observed that due to high running costs, not much desulphurization equipment is put into constant operation. However, if the amount of credit is determined by the actual desulphurization volume, this would no longer be an issue.

The other major advantage of sulphur tax is that there is almost no problem of pollution leakage that results when enterprises evade regulations by relocating to other areas. By taxing at the coal mines, the whole society would be regulated. Earlier we discussed the fact that China adopted the so-called 'Two Control Zones' policy to regulate key regions. Considering personnel and funding constraints, it might be the right decision to pool the necessary resources together and concentrate the enforcement efforts in the 'Two Control Zones' in order to increase the effectiveness of measures addressing environmental pollution. However, although emission from the 'Two Control Zones' accounts for 60 per cent of the national total, it by no means justifies implementing no measures for the remaining 40 per cent. If pollution sources evade regulation by moving to areas outside the zones, the issue of containing emission cannot be fundamentally addressed. With sulphur tax, on the one hand, limited policy resources could be concentrated to carry out centralized monitoring of those enterprises that are entitled to credits and, on the other hand, SMEs would be encouraged to act in full consideration of emission costs without monitoring.

Compared to command and control measures, because the sulphur tax could equalize the marginal cost of emission reduction by all polluters, an optimal balance between emission reduction cost and benefit could be achieved. However, since the tax rate is determined by the government, it is uncertain whether in the end it could turn into reality. More specifically, the government does not have sufficient information on polluters, particularly information on their emission reduction costs; so the final tax rate might be far from the optimal point and, thus, fail to achieve the best cost effectiveness. Lack of information is only one aspect of the difficulty in determining the optimal tax rate, as public opposition to the tax itself could also influence the process by which the tax rate is determined, something that in reality might result in incorrect decision-making. Generally speaking, the tax rate could be set at a lower level, which would be insufficient to bring about optimal emission reduction.

Despite the potential issue of whether an appropriate tax rate could be established, the sulphur tax has as its advantage the fact that the enforcement could be guaranteed to a great extent. This makes sulphur tax more desirable than other policy measures in light of current reality in China. In addition, the so-called 'Baumol-Oates Tax' could be effective in the Chinese sulphur tax system. Under the Baumol-Oates Tax framework, although it might be difficult to come up with the right rate in the very beginning, the government could start with a lower tax rate and make adjustments based upon the impact during implementation to ultimately arrive at an optimal tax rate (Ueta *et al.*, 1997; Morotomi, 2000). In other words, the tax rate could be adjusted by trial and error. Such an idea would face strong opposition from enterprises. However, in China, where the central government is politically very powerful, it is feasible to carry out adjustments to the tax rate in line with the Baumol-Oates Tax concept. The Chinese government decided to double and triple the emission fee from 2004–5, and this fully exhibits its power and authority to make adjustments to the tax rate. From the perspective of preparing the whole society for changes, it could turn out to be a more rational approach to start with a lower rate and gradually increase it to a reasonable level, so long as appropriate announcements are made to the public.

Alternative policy 2: emission credit trade system

So far, sulphur tax has not been listed as one of the policy options under study in China; in contrast, the emission credit trade system is being studied as a candidate among economic measures. Since 1999, State Environmental Protection Agency (SEPA) and the US Environmental Protection Agency (EPA) have worked on a joint project to study the feasibility of bringing an emission credit trade system into China. Case studies have been conducted in Benxi City, Liaoning, Nantong City in Jiangsu, and Taiyuan City in Shanxi province, and experiments on emission credit trading have been carried out in Nantong City. Although the above studies are still in the initial phase, breakthroughs are possible once the studies are brought into the next step to run pilot projects.

The emission credit trade system has its advantage in easier harmonization with the emission fee system. According to reports, an emission licence system has been implemented in 16 cities in China since 1991. Within three years, emission licences have been issued to 987 enterprises in 15 of the cities, and the system has effectively controlled 6,646 pollution sources, despite the fact that one city withdrew from the system. As of 2000, the necessary legal framework for the emission licence system had been established in 36 cities across 20 provinces (Wang *et al.*, 2002, p. 66). At this stage, the emission licence system is implemented as one of the ways to achieve total emission control, and no trading among pollution sources is allowed. However, the emission licence system has laid a certain foundation for bringing in the emission credit trade system.

Similar to sulphur tax, the emission credit trade system helps to equalize the marginal emission reduction costs of all polluters by making the emission credit trading price equal to the marginal cost of the whole society. The government only needs to set the initial amount of the emission licence and the market mechanism will automatically adjust the price to the optimal level through the trading system. As the number of emission licences in circulation is directly linked to emission volume, compared to sulphur tax, such a system could better control the emission volume. Therefore, in terms of preventing government failure, an emission credit trade system is a better option than a sulphur tax (Ueta *et al.*, 1997, Ch. 8; Field, 2000, Ch. 13).

However, similar to the emission fee system, the emission credit trade system also faces great challenges in achieving effectiveness in the course of its policy enforcement. The prerequisite of the emission credit system is to carry out strict monitoring on whether the participants observe the emission quota allowed by the emission licenses. This applies to both the buyer and the seller. Without effective and accurate monitoring, some parties might not purchase emission credit even though their emissions have exceeded their allowed emission quota and some might even claim to have extra credit and sell it on the market. Market credibility, which is the cornerstone of emission credit trade system, would face a severe challenge if such cases emerge one after another. In addition, it is necessary to shorten the intervals of emission volume monitoring in order to create more trading opportunities in the marketplace; hence, it is imperative for trading participants to install CEMs if the emission credit trade system is to be implemented. However, the installation cost of each set of CEM equipment amounts to 1.06 million yuan (Wang *et al.*, 2002, p. 88), and so presumably only a limited number of enterprises of a certain scale, such as power plants, could afford to install such equipment. Therefore, for medium and small pollution sources, such as industrial and residential boilers, little progress could be made in promoting the emission credit trade system.

In spite of the limitation of its effects on large pollution sources, specifically power plants, the emission credit trade system has certain effects. It has been mentioned earlier that the Chinese government in the future may require most power plants to install FGD but, according to cost calculation by the author, massive investment is needed to promote this policy. This might lead to worsening of cost effectiveness. Meanwhile, as shown in Figures 5.5 and 5.6, not all FGD projects enjoy the same cost effectiveness; instead, it varies greatly in line with the varying conditions of the power plants. Analysis of the distribution of coal resources from the perspective of quality shows that the sulphur content of coal, even raw coal, is very low in northern China, such as in Shanxi province. If power plants are located in those areas, they might reach the emission standard even if raw coal is consumed; hence, such power plants would suffer from very poor cost effectiveness if they were to install FGD equipment. Under the emission credit trade system, such power plants could purchase certain credits to meet the emission standard, which in practice amounts to partially sharing the costs of FGD equipment installed by other power plants. In other words, investment in FGD equipment by enterprises located in higher pollution intensity areas could be partially shared by other parties. This shows that there is great potential to increase cost effectiveness through the emission credit trade system.

Across the world, there is no precedent for combining the emission credit trade system and the environmental tax. However, in theory the two could be implemented at the same time. The problem is that to introduce an emission credit trade system, there must be monitoring facilities in place, which makes it rather difficult to incorporate the small and medium polluters into the system due to cost considerations. Hence, other policies need to be devised to address the small and medium sector.[8] For this, sulphur tax has the advantage. Therefore, massive synergy effects could be expected if the sulphur tax and emission credit trade system could be combined.

Alternative policy 3: enhancing the market mechanism in the coal market and setting up public coal trading companies

In this section, we discuss the importance of deploying clean energies, including CCT, and reflecting external environmental costs in pricing coal, particularly raw coal. Sulphur tax and the emission credit trade system should be instituted as specific measures to address the issue. However, unless coal is distributed based upon the market mechanism, it would make little sense even if external environment costs were reflected in coal prices through sulphur tax and the emission credit trade system. The market mechanism is essential because, under the sulphur tax or emission credit trade system, the users should be able to compare various options, for example, paying a higher transportation fee for low-sulphur coal or applying end-of-pipe technologies, such as FGD, and choose options based upon each specific situation, in keeping with the principle of achieving the best balance between cost and benefit. All these depend on a free market system for the users to compare and choose coal from various coal mines across the country.

Although many restrictions on coal mines and users have been eliminated, the current coal distribution system in China has a dual structure: large coal mines (key state coal mines) that supply to large enterprises and small to medium coal mines (town and village coal mines) that supply to local SMEs within the region. Many factors contributed to such a dual structure, including lack of scale of economy, transportation costs, insufficient market information and lack of intermediary functions in the marketplace (Horii, 2003). Unless this dual structure is discarded, sulphur tax and the emission credit trade system might not be able to fully play their roles as economic measures. Next we will explore how to address issues related to transportation cost and the lack of both information and intermediary functions.

Currently, long-range coal transportation depends on the railway, which is still state run, and the market mechanisms have little function. Consequently, compared to countries with similar geographic conditions, such as the US, transportation cost in China is on the higher end of the scale. In addition, rail transport quotas are not determined by the market mechanisms; instead, rail transport quotas are determined by political factors, similar to the traditional quota allocation approach under the planned economy. More than half of the annual rail transport quota for coal is allocated by the so-called Coal Ordering Conference, which is organized by the State Planning Commission (currently the National Development and Reform Commission) and attended by coal mines, coal users and transport departments, including the railway. Since only large coal mines and large enterprises are qualified to attend this conference, it institutionalizes one end of the dual structure of coal distribution in China.

Turning to the insufficiencies in both information and intermediary functions, these could be attributed to the fact that, in the past, coal was distributed under the state planning system and coal mines had no capacity for marketing on their own. This is particularly true for those small town and village coal mines, which account for a significant proportion of the national coal production. It is very difficult for small coal mines to know the demands of potential users located far a way. Since coal mines and consumers in China are both small in scale and geographically widely distributed, they both suffer from extremely high cost in terms of market information in coal trading. It is expected that someday there will be brokers who can match demand and supply; however, so far there is little progress in this regard.

If the dual distribution structure remains unchanged, then, due to transport constraints and lack of market information, users will still not be able to compare and choose coal based upon prices and quality, even if coal prices reflect the external environment costs. That will hamper the dissemination of clean energies, including CCT.

Under such circumstances, it is necessary to adopt certain measures to address current market defects and enhance the role of market mechanisms in China's coal market. More specifically, quotas for railway and port transport should be allocated through open bidding, and state involvement by the Ministry of Railways in the quota distribution process should be reduced, since it results in higher transportation costs. As the first step, transportation

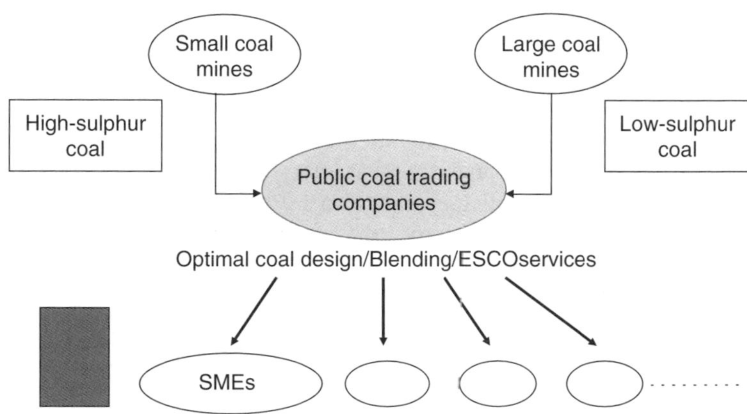

Figure 5.8   Benefits of public coal trading companies

capacity other than the quotas allocated at the annual Coal Ordering Conference should be distributed through open bidding, which should be open to all parties. To address the transportation cost issue associated with the dual coal distribution structure, it is necessary to carry out such reforms.

As a remedy for the other major issue hampering the market mechanism, the lack of brokers, the author would like to suggest public coal trading companies (Figure 5.8). Such companies not only engage in coal trading, but also provide CCT processing (coal washing, blending, briquette making and so on), consultation services on matters related to energy conservation, such as modification of the supplied coal based upon the boiler condition, and serve as an ESCO (Energy Service Company). The establishment of such companies would be beneficial not only for bringing down transaction costs but also for obtaining economy of scale in CCT processing and obtaining better understanding of the consumers' needs in terms of the promotion of CCT. It is a fact that the coal washing rate in small and medium township coal mines is extremely low. This is not only due to ineffective enforcement of environmental regulations but is also due to the fact that coal washing is unprofitable there because of the small scale of the coal mines in each town and village. Therefore, great cost effectiveness can be expected if public coal trading companies could carry out coal processing such as coal washing, blending, briquette making and so forth on a large scale. Furthermore, public coal trading companies could play an even more active role. If they provided ESCO services, for example, it would be possible for them to make profit by signing contracts with small and medium consumers to help with energy conservation because small and medium consumers have significant energy conservation potential but do not have financial resources to implement the energy conservation. Public coal trading companies could finance the energy conservation equipment cost and get financial rewards for the energy conservation achieved based on their ESCO contract.

Conclusion

In Sections 1 and 2, we critically reviewed current Chinese environmental policies, and, in Section 3, we proposed alternative policy options. It is true that there are some limitations on those alternative policy options. For example, sulphur tax has as its advantage the great effectiveness of the policy enforcement, i.e., since it can be collected at the coal mines, it markedly reduces the enforcement cost associated with monitoring and imposing penalties on violations. However, the limitation is that coal produced by illegal coal mines might be traded without paying the sulphur tax. There are similar potential limitations associated with the other alternative options proposed in this chapter. Such alternative options may be suboptimal options theoretically speaking, but considering their real-world effectiveness and feasibility of implementation, they are considered to be the best options. As far as sulphur tax is concerned, although it doesn't solve all the problems, it deserves serious consideration because collection of sulphur tax at the coal mines can better prevent evasion of the regulation and greatly alleviate coal-related air pollution to a greater degree than the emission fee system in which fees are collected from the end users.

In this section, the author will explain and comment on what makes the views expressed here different from other studies and summarize the analysis to draw implications that could benefit other developing countries in addressing air pollution.

The assumption that economic measures are the best solution, which used to be common, has already been abandoned by those who have conducted their studies based on Chinese realities (McElroy *et al.*, 1998; World Bank, 1997, 2001). While giving credit to the efforts made by Chinese government so far, they also commented positively on the policy mix combining command and control measures and economic measures. The primary difference between this study and the other studies is the views on issues related to managing pollutants from small polluters. Although the previous studies recognized the importance of addressing small polluters (particularly since small polluters present more risk to people's health as the chimneys are nearer to people), neither detailed analyses nor recommended policy options have been proposed in the other studies.

There is also a difference in opinions on the so-called 'Two Control Zones' policy, which targets cities suffering from severe air pollution. The other studies were published a few years ago when no specifics of the 'Two Control Zones' policy had been confirmed, and those studies basically approve of the idea of concentrating efforts to address air pollution in defined areas (World Bank, 2001). However in this study, based upon reviews on the implementation of the 'Two Control Zones' policy, the author evaluates the policy as undesirable because it induces evasion of the regulation and cause pollution leakage to the suburbs.

Ultimately, the differences in opinion originate from different ideas on command and control measures and economic measures, or the policy mix in environmental policy.

The other studies and this chapter all recognized problems related to the enforcement of environmental policies that stem from Chinese realities. In particular, all agreed that the local Environmental Protection Bureaus suffer from constraints on their policy enforcement capacity. To a certain degree, in light of the current low enforcement capacity, the previous studies had virtually given up on efforts directed at small polluters due to the associated high monitoring costs. Given the low enforcement capacity, from the perspective of cost effectiveness, it is reasonable to target large polluters as a priority since this could bring considerable results in reducing emissions.

However, as indicated in the Introduction, to a great degree, remarkable reductions in air pollution over the last few years can be attributed to command and control measures aimed at small polluters. The problem is that such command and control measures aimed at small polluters can be effective only so long as strong momentum is maintained in policy implementation. In the long run, once there is any loosening of policy initiatives, those small enterprises that had been shut down might secretly resume operations again, making the regulations mere pieces of paper at that time. In fact, the drastic increase of SO_2 emission in 2003 in China might be partially explained by the fact that some small polluters had started their operations again, stimulated by the favourable economic conditions.[9] In this way, no matter what the case may be in other countries, small polluters in China contribute considerably to the total emission. This fact by no means should be underestimated. However, the government is facing the dilemma of how to maintain the momentum and sustainability of command and control measures for small polluters.

The fact that the other studies gave positive comments to the 'Two Control Zones' policy could derive from their understanding of the limited enforcement capacity of the local Environmental Protection Bureaus. Given the limited enforcement capacity, the policy might be a realistic choice to ensure the effectiveness of the environmental policy. However, in this paper we have illustrated that it would be effective to reduce policy enforcement costs by introducing sulphur tax as one part of an environmental tax system. In forming environmental policies in China, it is very important to address evasion of regulations, and in this chapter it is shown that a sulphur tax system could address this issue effectively. The other studies were cautious in their review of economic measures. In their opinion, in China, there is not yet a market mechanism that smoothly functions. However, as illustrated in this chapter, from the perspective of policy enforcement costs, it does not make much sense to differentiate whether a policy should be called a command and control measure or an economic measure, and it is very difficult to judge which one is better than the other.

Through this chapter's analysis, it has been shown that command and control measures indeed suffer from poor cost effectiveness. On the other hand, economic measures also suffer, for example, in the case of the emission fee, from poor results in emission reduction when it is actually enforced, and quite possibly it may also suffer from failure to achieve the theoretical cost effectiveness. If technical requirements, which are one form of command and control measures, are adopted, it could help to reduce the enforcement costs. However, if emission standards, the other form of command and control measures, are adopted, it will lead to a repeat of the failures of many previous economic measures and increase the policy enforcement costs, such as monitoring costs. On the other hand, although environmental tax is a form of economic measure, it has the potential to reduce the policy enforcement costs. Therefore, depending upon the actual cases, one can expect different answers as to whether command and control measures or economic measures should be adopted.

In reality, almost all the policies consist of combined measures, although such policy mixes are hardly the best for achieving cost effectiveness. No doubt, cost effectiveness is a very important factor. However, it should not be viewed as the only criterion in evaluating policy framework. To ensure policy effectiveness, a policy framework combining multiple policy measures, as discussed in Section 3 of this chapter, should be established.

Sulphur tax is an effective measure for addressing air pollution in China. Considering the fact that market mechanisms, which are a precondition for economic measures such as sulphur tax to be fully effective, are still immature, the author proposes that the enhancement of market functions should be considered as one aspect of the environmental policies. The author points out that although the Chinese coal industry has been progressing towards a market economy, coal distribution in China remains constrained by non-market factors. Therefore, while introducing economic measures in environmental policies, it is also necessary to adopt measures to further promote the development of the market economy to ensure smooth functioning of the market mechanisms. On the other hand, to solely depend upon market mechanisms might result in the imperfect functioning of the market mechanisms due to such factors as trading costs. Depending upon the actual situation, it might be necessary to take such measures as setting up public coal trading companies, as suggested in this chapter. The author believes that all parties should be involved and that various policy mixes should be formed and should include some measures that were viewed as being outside the range of environmental policies in the past, to address this issue.

In fact, our discussion on solutions to air pollution in China could be applied to other developing countries when forming environmental policies.

First, since many developing countries suffer from such factors as lack of enforcement capacity, high monitoring costs and widespread policy evasion, it is important to design effective measures that take these factors into consideration.

Second, in many cases, small and medium enterprises are a major source of emissions in many developing countries. Furthermore, such SMEs often tend to be in the informal sectors. In order to manage emissions from SMEs, particularly in order to address air pollution, collection of environmental tax upstream from SMEs could be an effective policy measure that would help to reduce the monitoring costs.

Third, to ensure full enforcement of environmental policy in reality, in addition to cost effectiveness, time effectiveness is another essential factor to be considered. Therefore, the ultimate policy should be a combination of various measures.

Fourth, an environmental tax system is an effective policy measure for developing countries; however, because an environmental tax system is an economic measure, positive achievements are possible only if the market mechanisms are functioning. The problem in many cases is that, since developing countries are still in the transitional period, they often suffer from market distortions. Therefore, measures that could help to improve the market function or enhance the market economy should be considered as part of the environmental policies.[10]

Notes

1. SO_2 emission in 2003 was 21.6 million tons, up by 12 per cent compared to the previous year. Since energy consumption in 2003 had increased by 13 per cent during that time, the increase in SO_2 emission is understandable. However, if we take a look at the figures for 2002, we will see that although energy consumption grew by 10 per cent compared to the previous year, SO_2 emission was lower than the previous year. Since more attention was paid to environmental protection in 2003, it is rather odd that energy consumption increased both in 2002 and 2003, and SO_2 emission dropped in 2002 but increased considerably in 2003. Therefore, a more reasonable explanation could be that there were errors in the statistics up to 2002. Even so, I agree that China has made considerable progress in addressing air pollution, as figures released in 2003 show that, from 1995–2003, the energy consumption grew by 28 per cent and the SO_2 emission was lowered by 9 per cent during the same period.
2. Evasion of regulations is widespread in China and is not limited to the environmental protection sector. Therefore, special attention should be given to how to prevent such evasion when forming policy measures. Although the single largest factor that has contributed to the alleviation of air pollution in China since 1998 is the shrinking of coal consumption, there is speculation as to whether there have been any errors in the statistics due to difficulty in obtaining figures on coal consumption behind the scenes. In Horii (2001) and Akimoto, Ohara, Kurokawa and Horii (2007), the authors concluded that there could be around 150–200 million tons omitted from the statistics on coal consumption for 2000. Since emission of air pollutants in China is calculated based upon the consumption volume of fuel, we should maintain a certain degree of alertness with regard to the statistical reduction of emissions as mentioned above. However, considering that upto 2002, on the one hand, coal consumption statistics have kept lower than 1997 levels and, on the other hand, it seems that certain progress has been made in curbing air pollution

thanks to the implementation of the command and control measures as described in this chapter, the author concludes that the current emission reduction efforts have produced certain results. In contrast, Otsuka (2002) has described various specific instances of evasion of regulations.

3. The figures cited here only cover the initial investment, excluding operational costs. However, as shown in Table 5.1, the operational costs of FGD should by no means be neglected. If operational costs were to be considered, even for simplified desulphurization equipment, the cost to performance would be very poor. This also applies to CFBC, since it also involves significant operational costs. The figures cited in this chapter excluded such considerations due to difficulty in collecting exact operational costs.

4. Urban natural gas projects have certain advantages since it is mainly used for cooking and heating purposes, where it has a large impact on people's health and reduces indoor pollution. It is very effective in reducing hazardous emission. The author fully realizes that we should not simply view it from the single perspective of the relationship between investment and SO_2 emission; instead, considerations should be made from a broader perspective.

5. The information on collection of emission fees in this chapter is based upon a field survey conducted by the author in Henan, Anhui, Hunan and Sichuan provinces in 2002 and 2003. The information released by the State Administration of Environmental Protection (2003b) also revealed the same issues (pp. 35, 40, 49, 51, 52).

6. In response to the issue of overcharging polluters by the local Environmental Protection Bureaus for the purpose of financing their own budgets, as of 2002, the emission fee is paid through bank transfers and invoices were made mandatory. The payment accounts are no longer managed by the Environmental Protection Bureaus but by Finance Bureaus. In the future, part of the emission fee revenue shall be allotted to the environmental protection departments on the county level at proportional rates. However, since such reforms will directly result in lower budgeting for the Environmental Protection Bureaus on the county level, which are responsible for implementing the relevant environmental laws and regulations, it is worrisome that their insufficient law enforcement capacity may be further weakened.

7. Furthermore, it is a popular view among the local law enforcement authorities that they have done their job so long as the environmental polices have been implemented within the target zones. In some cities, efforts have been made to encourage enterprises within the target zone to relocate to the suburbs. For example, since Taiyuan City of Shanxi province suffers from severe air pollution, the whole city was included in the 'Two Control Zones' area; recently, however, many polluting enterprises have been relocated from Taiyuan to neighbouring cities, and it is said that the Taiyuan City Government encourages such practices. In Henan and Hunan provinces, the author has also heard the local officials express the same view, which they took for granted.

The problem that SO_2 emissions cause is damage to the local forests and to people's health. Hence, some people held the view that policies should focus on reducing the concentration of emissions in densely populated areas, instead of focusing on reducing the emission volume. From this perspective, the 'Two Control Zones' policy has certain merits since it helped to relocate the polluters from the urban areas to the suburbs. This certainly helps to reduce the concentration of emissions. The problem is whether the areas outside the 'Two Control Zones' are really sparsely populated. With the exception of megalopolises like Beijing, Shanghai and a few other coastal cities, for some inland cities included in the 'Two Control Zones', the population density in the urban area is not much higher than in the suburban area. In China, due to restrictions of the household registration system, people could not freely migrate to cities; hence the concentration of population in the cities is not as obvious as in Japan. The 'Two Control Zones' only cover the urban area and exclude the neighbouring suburbs. Due to such considerations as infrastructure and utility services, it is rather natural for the polluting enterprises to choose to relocate to the suburbs, which are so close to the urban area, although population density in the suburbs might not be much lower than in the city proper. Considering the massive emission volume in China, the issue of pollution leakage to areas outside the 'Two Control Zones' should by no means be neglected.

8. The American acid rain prevention programme (Clean Air Act Amendments: CAAA) has been widely promoted as a successful example of emission credit trading. In the programme, power plants were the primary participants. In the US, the power industry accounts for 91.2 per cent of the total coal consumption in the country; hence, being the only primary coal consumer, their participation in the programme could achieve remarkable results in terms of environmental protection. On the other hand, although the power industry is also the largest coal consumer in China, it only accounts for 45 per cent of the total coal consumption, the remaining being consumed by small and medium industrial and residential boilers. Therefore, emission reduction through the emission credit trade system would be limited in China and more thought should be given to the great differences between the coal consumption structure in the US and in China.

9. As indicated in Note 1, there were inaccuracies in statistics on SO_2 emission in 2002; hence, compared to the unrealistically low statistics in 2002 and before, the figure grew drastically in 2003. While pursuing the causes for the under-reporting of 2002 statistics, it is appropriate to conclude that the effects of compulsory shutting down of small polluters through command and control measures have not been as great as reported. If so, the difference between the figures of 2002 and 2003 reveals the scale of SO_2 emissions by small polluters, and this shows that the emission by small polluters should by no means be neglected.

10. If public coal trading companies are to be established, relevant parties need to be on guard against monopoly. However, under the current situation, apparently enterprises equipped with functions described in this chapter could hardly emerge from the private sector. In the past, under the planned economic system, there were the so-called Fuel Companies under the former Ministry of Commerce, and those companies were supposed to serve as public companies and provide fuel to small and medium boilers in urban areas. Since the reforms to the economic system, such companies are almost all bankrupt, since they were not able to reform their system of living on government allowances. Here the author would like to stress that there are fundamental differences between the public coal trading companies proposed in this chapter and the Fuel Companies in the past. From the very beginning, such public coal trading companies should be established as market-oriented companies to be privatized later on. Unlike the former Fuel Companies that failed to produce any added value and only bought coal from one party and sold to another party, the new public coal trading companies should be able to serve as ESCO, which help their clients to conserve energy by providing consultations and suggestions and, thus, help enterprises to obtain direct benefits. It might be conducive to clarify the schedule of each step for the establishment of such public coal trading companies and publicize the timetable for privatization.

References

Akimoto, Hajime, Toshimasa Ohara, Jun-ichi Kurokawa, and Nobuhiro Horii (2007) 'Verification of Energy Consumption in China during 1996–2003 by Satellite Observation', *Atmospheric Environment, Elsevier Science Publishers.*

Blackman, A. and W. Harrington (1999) 'The Use of Economic Incentives in Developing Countries: Lessons from International Experience with Industrial Air Pollution' *Resources for the Future Discussion Paper* 99–39, Washington, DC, Resources for the Future.

Chen Qingru (2001) '21 shiji de jiejing meitan nengyuan' (Clean Coal Energy in 21st Century). *Jiejing meitan jishu (Clean Coal Technology).* Clean Coal Engineering and Research Centre, China Coal Research Institute, 7 Special Issue Oct. pp. 11–14 (in Chinese).

Chen Wenmin, Zizhao Chen and Huaizhen Chen (1999) *Dongli peimei (Coal Blending for Steam Coal).* Beijing, Coal Industry Publishing House (in Chinese).

Fang, J., Zeng T., Lynn I. Yang S., Oye K. A., Sarofim A. F., and J. M. Beer (1999) 'Coal Utilization in industrial boilers in China – a prospect for mitigating CO_2 emissions'. *Applied Energy* 63, pp. 35–52.

Field, Barry C. (2002) *Kankyo-keizaigaku nyumon (Environmental Economics; An Introduction).* Tokyo, Nihon-Hyouron-sya (in Japanese).

Former Ministry of Coal Industry (eds.) (various years) *Zhongguo meitan gongye nianjian (China Coal Industry Yearbook).* Beijing, China Coal Industry Publishing House (in Chinese).

Fujikura, Ryo (2002) 'Nihon no chihou-kokyodantai no Iou-sankabutsu-taisaku: kodo-keizai-seichoki ni jissisareta kogai-boshi-kyotei to gyosei-shido' (Local Measures in Japan for Controlling Sulfur Oxide Emissions: Pollution Control Agreements and Administrative Guidance Adopted during the Era of Rapid Economic Development), in Terao, T., and Otsuka, K. (eds), *'Kaihatsu to kankyo' no seisaku-katei to dainamizumu: nihon no keiken, higashi-ajia no kadai (Dynamism of the Policy Process in 'Development and the Environment': The Experiences of Japan and Problems in East Asia).* Chiba: Institute of Developing Economies (in Japanese).

Horii, Nobuhiro (2000) 'Sekitan sangyo: sangyo-seisaku niyoru shigen-hozen to jizokuteki-hatten' (Coal Industry: Resource Reservation and Sustainable Development through Industrial Policy). In Marukawa Tomoo (eds.) *Ikoki chugoku no sangyo-seisaku (China's Industrial Policys in Transition).* Tokyo, IDE, pp. 203–46 (in Japanese).

Horii, Nobuhiro (2001) 'Chugoku ni okeru enerugi shohi gekigen no haikei' (What's Behind the Dramatic Reduction of Energy Consumption in China) *Coal Journal,* (42), CCUJ, Tokyo, pp. 3–7 (in Japanese).

Horii, Nobuhiro (2003) 'Chugoku sekitan-sangyo no sijo-togo heno sogai yoin' (Impediments to Integration of China's Coal Market: Questionnaire Survey on Coal Distribution Structure in Hunan Province), Azia Keizai: Institute of Developing Economies IDE-JETRO, Chiba, pp. 2–37 (in Japanese).

Li, Zhidong (1999) *Chugoku no kankyo-hogo sisutemu (Chinese Environmental Protection System).* Tokyo, Toyo-keizai-sinpousya (in Japanese).

Liu, Maochun (2000) *Ranmei gongye guolu jieneng shiyong jishu (Energy Conservation Technologies for Coal Fired Industrial Boilers).* Beijing, China Power Publishing House (in Chinese).

Matsuoka, Shunji (2000) 'Tojokoku ni okeru kankyo-seisaku no koritsuteki jissi toha nanika?: kankyo-seisaku syosyudan to koritsusei' (What is the Effective Enforcement of Environmental Policy in Developing Countries? Various Measures of Environmental

Policy and its Efficiency), *Journal of International Development Studies* 9, (2), The Japan Society for International Development, pp. 17–37 (in Japanese).

McElroy Michael B., Chris P. Nielsen and Peter Lydon (1998) *Energizing China: Reconciling Environmental Protection and Economic Growth.* Harvard University Press.

Morotomi, Toru (2000) *Kankyo-zei no riron to jissai (Environmental Tax: Theory and Reality).* Tokyo, Yuhikaku (in Japanese).

National Bureau of Statistics of China. (Various years) *Zhongguo tongji nianjian (China Statistical yearbook).* Beijing, China Statistics Press (in Chinese).

OECD (2002) *Kankyo-kanren-zeisei: sono hyoka to donyu senryaku (Environmentally Related Taxes in OECD Countries: Issues and Strategy),* Tokyo, Yuhikaku (in Japanese).

Otsuka, Kenji (2002) 'Chugoku ni okeru kogyo-osengen-kisei no jissi-katei: 1990 nendai-kohan iko no kisei-seisaku no jikkosei to sono joken' (Implementation Process of Regulations for Point Sources of Industrial Pollutants in China: Effectiveness and Conditions of Regulatory Policy since the Late 1990s) in T. Tadayoshi and K. Otsuka (eds), *'Kaihatsu to kankyo' no seisaku-katei to dainamizumu: nihon no keiken, higashi ajia no kadai (Dynamism of the Policy Process in 'Development and the Environment': The Experiences of Japan, Problems in East Asia.)* Chiba, IDE.JETRO, pp. 139–85) (in Japanese).

State Environmental Protection Agency (1997) *Zhong, xiao xing ranmei guolu yanqi chuchen, tuoliu shiyong jishu zhinan (Removal Technologies of Dust and Sulphur from Flue Gas of Medium and Small Coal Fired Boilers).* Beijing, China Environmental Science Publishing Company, (in Chinese).

State Environmental Protection Agency (2003a) *Zhongguo huanjing tongji nianbao (China Environment Statistics Annual 2002).* SEPA (in Chinese).

State Environmental Protection Agency (2003b) *Paiwu shofei zhidu (Emission Fee System).* Beijing, China Environmental Science Publishing Company, (in Chinese).

Sterner, T. (2003) *Policy Instruments for Environmental and Natural Resource Management,* Washington, DC, Resources for the Future.

Tsutsumi, Kenzo, Tomohiko Iwasaki and Syodai Tsukahara (2003) *Enerugi manejimento: ESCO, ESP no choryu (Energy management: ESCO and ESP Trend).* Newspaper Department, the Electric Association of Japan (in Japanese).

Ueta, Kazuhiro, Toshiriro Oka, and Hidenori Niizawa (1997) *Kankyo-seisaku no keizaigaku: riron to genjitsu (Economics of Environmental Policy: Theory and Reality).* Tokyo, Nihon-Hyouron-sya (in Japanese).

Wang Jinnan, Yang Jintian, Stephanie Benkovic Grumet, Jeremy Schreifels, and Ma Zhong (eds) (2002) *Feasibility of SO_2 Emission Trading into China,* Beijing, China Environmental Science Publishing Company (in Chinese).

World Bank (1994) *Pre-Feasibility Study on High Efficiency Industrial Boilers,* Subreport Number 11, Washington, DC, World Bank.

World Bank (1997) *Clear Water, Blue Skies,* Washington, DC, World Bank.

World Bank (2001) China Air, Land, and Water: Environmental Priorities for a New Millennium. Washington, DC, World Bank.

Zhou Dadi *et al.* (2003) *2020 zhongguo kechixu nengyuan qingjing (Scenario for China's Sustainable Energy to 2020).* Beijing, China Environmental Science Publishing Company (in Chinese).

6
Rating Programme Revisited: In the Case of Indonesia

Michikazu Kojima

Introduction

The rating programme for companies in Indonesia is regarded as a model of environmental policy from the viewpoint of environmental management in developing countries. PROPER PROKASIH, a rating programme concerned with water pollution control in Indonesia, attracts international attention because the programme is an alternative environmental policy based on disclosure of information.[1] In addition to the same kind of programme being carried out in the Philippines, India and China, similar rating programmes are also planned in Mexico and Colombia.[2]

PROPER PROKASIH also attracted the attention of researchers. The programme was taken up by World Bank (2000), Afsah *et al.* (2000) and López *et al.* (2004). World Bank (2000) pointed out that the efforts of companies to make environmental measurements increased because the rating programme stipulates that companies be aware of local residents who demand environmental measurements and the evaluation of the company in the stock market. Afsah *et al.* (2000) analysed PROPER PROKASIH by sending a questionnaire to managers of companies. López *et al.* (2004) analysed panel data of factories targeted in PROPER PROKASIH. Their analysis shows that the programme reduces the amount of effluent from the factories. However, the objectives of these research studies were focused on the impact of PROPER PROKASIH, and they neglected the historical background of pollution regulation in Indonesia. Moreover, the meaning and the limitations of the rating programme were not clearly identified in the context of environmental management in Indonesia. For example, prior to PROPER PROKASIH, ratings were released in PROKASIH (Clean River Program) in Jakarta in 1991, and this marked the beginning of the rating programme in Indonesia. This first experience may have had impact on the behaviour of companies targeted in later ratings, but previous studies have neglected it.

On the other hand, Rock (2002) reviews the history of establishment of agencies concerned with the environment, such as the Environmental Impact

Management Agency, and programmes such as PROPER PROKASIH and PROKASIH. The study shows that the personnel in charge of the environment introduced environmental policies such as PROPER PROKASIH creatively under the Suharto administration, which put national priority on 'development'.

In the period of economic crisis and political turmoil from 1997–2002, PROPER PROKASIH was terminated. However, it was considered one of the successful environmental policies in Indonesia. In 2002, the Ministry of the Environment started the preparation of the PROPER programme, which is an expanded version of PROPER PROKASIH. PROPER includes ratings not only for water pollution control but also for air pollution control, hazardous waste management, environmental management systems and so forth.

In this chapter, while looking back upon the history of measurement of factory effluents in Indonesia, the meaning and the limitations of the rating programmes are examined. The number of companies involved in the programmes and the substances to be reported are also examined, thereby revealing the limitations of the programmes. Furthermore, the impact of the new rating programme, PROPER, which succeeded PROPER PROKASIH, is discussed.

First, in Section 1, the establishment of the ministries and government offices specializing in the environment and environmental regulation is outlined in order to provide background to understand the meaning and limitations of rating programmes in Indonesia. Section 2 outlines PROKASIH, PROPER PROKASIH and PROPER, respectively. The meaning and limitations of these programmes are discussed in Section 3.

1.   Development of organizations and regulations in the 1970s and 1980s

Organizations in central government

In the days when Indonesia was a colony of the Netherlands, some people were concerned about the protection of natural resources in Indonesia.[3] After independence, the Indonesian government came to be conscious of environmental problems around the time the United Nations Conference of the Human Environment was held in Stockholm in 1972. It was also a time when the number of factories was increasing gradually following the enactment of the Foreign Investment Act in 1967 in Indonesia.

A decree concerning prevention of atmospheric, water and sea pollution in the city of Jakarta was issued as a city decree in 1971. It forbade dumping wastewater in rivers, swamps, etc. The Environment and Resource Section was established in the National Development Planning Agency (*Badan Perencanaan Pembanguan Nasional* [BAPPENAS]) in 1972. As a position specializing in the environment, the position of State Minister for Development and the Environment was created in 1978, and this position was taken over by the State Minister for Population and Environment in 1983. Although the office supporting this minister is often called 'Ministry', it had few staff members compared with other ministries and did not have local branch offices. Its status was different from other ministries such as the Ministry of Energy and Mines and the Ministry of Forestry, which have many staff members and local offices. In the Indonesian language, the term meaning *Kantor Menteri* was used for office of State Minister for the Environment, while *departmen* was used for the other ministries.

The State Minister for the Environment was in charge of coordination between governmental agencies on environmental issues, but this minister did not have authority to enforce environmental regulations. Enforcement of environmental regulations was an obligation of ministries such as the Ministry of Industry, Ministry of Energy and Mining, and Ministry of Forestry, as well as of their provincial offices and so forth. However, the enforcement of environmental regulations was weak in the 1970s and 1980s.

The above is the background to the establishment in June 1990 of the Environmental Impact Management Agency (*Badan Pengendalian Dampak Lingkungan* [BAPEDAL]), which was in charge of enforcement of environmental regulations, such as inspection of companies. However, the authority of this agency was still restricted. The Environmental Impact Management Agency and the Office of State Minister for the Environment were unified in 2002, and the Ministry of the Environment was launched.

Basic environmental law and environmental regulation

From the middle of the 1970s, environmental regulation progressed gradually. Under the decree of the Minister of Health in 1977 (No.173/Men. Kes/Per/VII/77), the standard of the effluent discharged to swamps, lakes and rivers, etc., from industry, mining and households was defined.

The Act on Basic Provision for the Management of the Living Environment (Law No. 4 in 1982, hereinafter referred to as the Basic Environmental Act) was enacted in 1982. It states that the objectives of management of the living environment are to achieve harmonious relations between humans and the living environment, to control wisely the utilization of natural resources, to implement development with environmental consideration for the interests of present and future generations and so forth (Article 4 of the Act). Every person has the right to a good and healthy living environment and has the obligation to maintain the living environment and to prevent and abate environmental damage and pollution (Article 5). Mention is also made concerning the obligation of people in enterprise, to the effect that 'every person engaged in an enterprise has the obligation to maintain the continuing capability of the harmonious and balanced living environment to support continued development' (Article 7[1]).

The Industrial Act enacted in 1984 stated, in the first clause of Article 21, that it is necessary to care about the environment and that companies bear responsibility for preventing environmental destruction and contamination. The second clause of Article 21 states that the government is to perform instruction and regulation, etc., for the prevention of contamination. However, in the third clause of Article 21, certain industries classified as small-scale industries are exempted from the first clause of Article 21.

Governmental Decree No. 17 in 1987 was concerned with promotion and regulation of industrial development by the authorities and, although the Minister for Industry was given authority for industrial regulation including pollution control and promotion of industries, some parts of industries were managed by other ministers. The Ministry of Agriculture was in charge of environmental management in the pharmaceutical industry, and the Ministry of Energy and Mining was in charge of environmental management in the industries of oil refining, natural gas and processing of non-ferrous-metal minerals, etc. The government offices supervising specific industries are also in charge of environmental management of those industries.

Under Governmental Decree No. 13 in 1981, which was concerned with permits for industrial enterprise, a duty was imposed on companies to make an effort to prevent environmental destruction and to control pollution caused by industrial activity (Article 14). Although Article 10 defined the conditions under which industrial licences would be cancelled, the article does not clearly mention environmental destruction or failure of pollution control among the reasons for cancellation.

The decree of the Minister for Industry in 1988 (No.134/M/SK/4/1988) listed industries that should perform environmental impact assessments. Moreover, the punishment for factories causing contamination intentionally or by negligence includes the stoppage of industrial activity.

The Ministry of Industry decided on the detailed procedure for factory licensing in Governmental Decree No. 13 in 1995. It was determined that an industrial licence could be cancelled when an industry caused environmental destruction or contamination that exceeded environmental standards. Moreover, small-scale industries in specific sectors were exempted from the requirement to acquire industrial permits. Industries that did not cause environmental destruction through their production process and did not overuse natural resources were specified by the decree of the Minister for Industry (No.146/M/SK/7/1995) in 1995.

In the Regulation Concerning Industrial Licences, Licence Extensions and Industrial Registration stipulated in the decree of Ministry of Industry and Trade in 1995, industrial licences are not required for companies whose total investment is less than 200 million rupiahs, excluding land and building. Moreover, the procedure was also defined for revoking licences from factories that cause environmental pollution. First, a written warning is sent to the factory that has caused contamination. If the factory does not improve even after written warnings have been sent three times at intervals of one month, then the industrial licence is suspended for six months. If the factory improves within six months, then the licence will reinstated; however, the licence will be revoked if there has been no improvement during the six-month suspension period. Using this measure against companies that cause environmental pollution, it takes a long time to stop the operation of industrial activity causing pollution.

A legal framework to control industrial pollution has been developed since the 1970s. While authority to stop operation of companies discharging pollutant was given to ministries which promote industrial activity, no authority was given to the State Minister for the Environment. It is very difficult for the ministries to take action to close a factory that is causing contamination.

Enforcement

Although the authority to stop the operation of factories causing pollution was given to central ministries or local governments, this authority was not often exerted. According to Braadbaart (1995), the Ministry of Industry and the government employees at the provincial level in West Java did not understand anti-pollution measure technology. They also could neither give advice nor evaluate the efforts of the factories. Furthermore, factory personnel reported that government officials sometimes required bribes.[4] Even if they saw the government report on effluent in the 1990s, many companies were not meeting the effluent standard.

As a result of neglecting contamination, some factories faced intense disputes over pollution. In an extraordinary case, the surrounding residents set a factory on fire. Anti-pollution measures gradually grew stronger, starting in the first half of the 1980s in such areas.[5] Pargal and Wheeler (1996) verified statistically whether the effluent level of factories decreased with surrounding residents' pressure, using data from monitoring during 1989–90, before governmental regulation got into its stride. It is shown clearly that effluent levels fall in areas of higher income or higher academic level, and that effluent levels of state-owned firms are higher than other companies. The effluent level of foreign-affiliated firms were not significantly different from those of local private enterprises.

2. Rating programs in Indonesia

After the appreciation of the Japanese yen and newly industrializing countries' (NICs') currencies in the latter half of the 1980s, foreign direct investment into Indonesia increased. Pollution became a huge social problem in this period. The first boycott movement against manufacturing goods was organized by environmental NGOs in 1991, in response to the pollution in Semarang.

The basic environmental regulations had been developed in the 1980s. However, enforcement of the environmental regulations was weak. How to strengthen the enforcement of environmental regulations under limited authority was a priority of the Office of the State Minister of the Environment.

PROKASIH

PROKASIH (*Program Kali Bersih* [Clean River Program]) is the starting point of effective government regulation of water pollution. Although PROKASIH was initiated by the proposal of the Office of State Minister for the Environment in 1989, it was provincial government that determined target rivers and target factories. At the beginning, installation of pollution control measures was urged for factories located in the watershed of 20 rivers in eight provinces. The number of target rivers, target factories and types of industries were increased. Seventeen provinces, 37 basin systems and 77 rivers become targets in FY1999/2000. The Office of State Minister for the Environment and Environmental Impact Management Agency provided technological assistance and supported information collection.

The target factories are forced to sign an agreement with the provincial government about the schedule to satisfy effluent standards. The provincial government monitors whether the target factory is following the agreement, and the Office of the State Minister for the Environment evaluates the activity of each provincial government. In addition to instituting countermeasures against pollution by factories, the State Minister for the Environment travelled down the Ciliwung River, which flows into Jakarta, by boat in order to raise awareness concerning cleaning the river.

By the governmental decree concerning Control of Water Contamination (Governmental Decree No. 20 in 1990) in June 1990, it was stipulated that the State Minister for the Environment is to determine the effluent standard after talking with other related ministers, etc. (Article 15[1]), and water quality control is to be enforced by the provincial governor. An organization specified by the provincial governor is to perform the enforcement by checking effluent treatment facilities and taking samples, etc.

The Office of State Minister for the Environment was motivated to institute PROKASIH because the provincial governments, which had the authority to control water pollution, were not fully functioning. Moreover, it was considered difficult to satisfactorily tighten up the regulation of various pollution sources under the existing regulations, and it was thought that PROKASIH could improve the enforcement capability of the local government step by step.

On 2 September 1990, an international workshop was held by the Office of State Minister for Population and the Environment in cooperation with the Canadian International Development Agency.[6] The workshop had 57 participants from the ministry, provinces which have participated in PROKASIH, donor organizations from eight countries that are assisting Indonesia, embassies, the UN Development Programme and the World Bank. The purpose of the meeting was to promote mutual understanding among the organizations that were involved or were planning to be involved in PROKASIH.

A chart was made in the preparatory process of the workshop in order to gain an overview of the situation of PROKASIH. Eight factors were listed that affected the performance of PROKASIH, such as 'river water control at the provincial level is not achieving the desired effect', 'effluent standards for each industry remain undecided' and 'lack of inventory of the water quality of a river, effluents, and monitoring'. Furthermore, the situation used as the background of the eight factors was analysed. Problems that were pointed out include the following: 'the river water pollution control at the provincial level is not achieving the desired effect', 'coordination with the government organization engaged in river water pollution control takes time', 'the contents of plans for river purification programmes at the provincial level are not suitable (i.e., selection of industries and targets)' and 'there is no concept of comprehensive watershed management', etc.

After confirming their understanding of the above problems, the workshop participants made a draft of guidelines for carrying out PROKASIH:

1. Inventory of river water quality and flow rate
2. Evaluation and determination of river use patterns and ambient water quality standards
3. Inventory of pollution sources
4. Determination of pollution carrying capacity
5. Identification and selection of industrial targets for river pollution control
6. Establishment of pollution licensing system
7. Monitoring of industrial waste at source
8. Monitoring of river water quality and flow rate
9. Development and application of incentives and disincentives
10. Establishment of pollution control guidance
11. Development of public participation

In (9) above, an 'award for effective wastewater management' was mentioned. Moreover, although there was no specific proposal, the importance of public participation was recognized in (11). This stance, which recognized the importance of public awareness and participation, was the basis of the rating programme.

A matrix summarizing activities of domestic organizations and donor agencies in above-mentioned field was also made. A comprehensive range of content was presented and examined in this workshop, and the workshop was able to form a starting point for composing a plan for the new project and an ex post evaluation of PROKASIH. However, almost no reports or papers have referred to this workshop.

Regarding effluent standards, which were a topic in this workshop, Decree No. 3 of the Ministry of Population and Environment was issued in 1991, and this decree stated that the effluent standards were to be determined according to the quantity of production of each industry. The target industries were caustic soda, electroplating, tanning, oil refining, palm oil, paper and pulp, rubber, sugarcane, tapioca starch, fibre, urea fertilizer, ethanol, monosodium glutamate and plywood. Following that, effluent standards for hospitals, supermarkets and cleaning shops were defined, and these industries were included among the target industries of PROKASIH.

As part of PROKASIH in Jakarta, the names of 95 companies and the amounts of pollution they emitted were released in October 1991 (Table 6.1). This was the beginning of the rating programme in Indonesia. Although 17 of these companies were meeting the effluent standard, 56 companies had not attained the effluent standard. The official announcement of these company names was taken up by many newspapers and magazines. The companies were urged to undertake anti-pollution investment. Confectionery company PT. Trebor, which was the worst polluter, was forced to stop operation for half a year and to invest 700 billion rupiah in pollution control measures from the beginning of 1992.[7] As a result, the biological oxygen demand (BOD) pollution load from surveyed factories decreased rapidly to 0.54 kg per day in 1993 from 294 kg per day in 1991.[8]

There was a reason why the State Minister of Population and Environment released the names of companies to the public. The State Minister of Population and Environment, as well as the Environmental Impact Management Agency, has weak authority. As mentioned above, provincial governors, the Minister of Industry and other ministers have authority to control pollution emitted by industry, such as by cancelling the operation permit of a factory. It can be concluded that the State Minister of Population and Environment decided to put companies that caused contamination under public pressure by releasing their names.

Moreover, a movement boycotting the products of a company polluting the water of the Tapak River in Semarang was organized by NGOs, half a year before the names of companies were released in Jakarta. It was the first trial

Table 6.1 Pollution control of target companies in PROKASIH Jakarta (October 1991)

Rank	Criteria	Number of companies
1	Operating wastewater treatment facility properly	17
2	Operating wastewater treatment facility, but not getting expected result	36
3	Testing wastewater treatment facility	2
4	Wastewater treatment facility was installed, but not operated	7
5	No wastewater treatment facility	11
6	Bankrupt or relocated factory	22

Source: Compiled from *Tempo*, 2 November 1991.

of boycotting in Indonesia for reasons of environmental pollution.[9] A mediation committee was established that consisted of representatives of Environmental Impact Management Agency, the city government, an NGO, the lawyer representing the residents and companies. It was agreed that the companies would make anti-pollution investments and pay compensation for the damage, etc. This boycott suggested the possibility that official announcements of polluting companies in Jakarta might lead to another boycott movement, and companies were afraid of such movements.

PROPER PROKASIH

PROPER PROKASIH, a programme that rated the companies in PROKASIH, was started in 1995. The names of companies and the degree of their countermeasures against pollution were released. This programme was supported partially by the World Bank. In the rating system, companies were classified into five categories such as gold, green, blue, red and black. While black companies did not perform anti-pollution measures at all, red companies made efforts to control pollution but did not reach the effluent standard (Table 6.2). The performance criteria covered the areas of air pollution control and hazardous waste management. However, the data on air pollution and hazardous waste management was limited, with only the gold and green companies being scrutinized in the area of air pollution, hazardous waste and so forth.[10] Ratings were based on the effluent data collected by questionnaires, monitoring for PROKASIH and information from citizens. PROPER PROKASIH is trying to urge companies to take more positive environmental

Table 6.2   Rating scheme of PROPER PROKASIH

Compliance status	Colour rating	Performance criteria
Compliant	Gold	All requirements of the green rating, plus similar pollution control for air, hazardous waste. Polluter achieves high international standards by making extensive use of clean technology, minimizing waste, preventing pollution, recycling and so on.
	Green	Pollution level is significantly lower than the effluent standard. Polluter also disposes of sludge properly, ensures good housekeeping, keeps accurate pollution records and maintains the wastewater treatment system.
	Blue	Polluter applies effort sufficient only to meet the standard.
Non-compliant	Red	Polluter makes some effort to control pollution, but not enough to sufficiently to achieve compliance.
	Black	Polluter makes no effort to control pollution, or causes serious environmental damage.

Sources: Compiled from PROPER-PROKASIH Team and PRDEI (1995) and World Bank (1997).

measures by releasing the names of companies that did not reach the effluent standards and which made efforts above the effluent standard.

The first rating was released in June 1995 and covered 187 companies located in Jawa, Sumatra and Kalimantan. The second one was released in December 1995, and the third one was released in October 1996.[11] The forth rating, which was evaluated in March 1997 and released in July 1997, covered 270 companies. Among the 270 companies evaluated in March 1997, 217 companies (80.4 per cent) were companies targeted in PROKASIH, 15 companies (5.5 per cent) participated PROPER PROKASIH voluntarily and 38 companies (14.1 per cent) were selected from companies that were not targeted in PROKASIH. Concerning the industries involved, 71 companies (26.3 per cent) were selected from the textile industry, 29 companies (10.8 per cent) from the paper pulp industry, 27 companies (10.0 per cent) from the plywood industry and 26 companies (9.6 per cent) from the rubber industry. Concerning ownership, 64 companies (23.7 per cent) were state-owned firms, 166 companies (61.5 per cent) were foreign capital firms and 40 companies (14.8 per cent) were domestic private firms.

Changes in the number of companies assigned to each colour rating from the first rating to the fourth rating are as shown in Table 6.3. There were no companies classified as gold at any time. The number of companies classified as green and blue increased. Looking only at the 187 companies that were rated all four times, there was a decrease from 64.7 per cent to 50.8 per cent in the percentage of companies that did not satisfy the effluent standard and were classified as red or black.

Table 6.4 summarizes the results of ratings from the second time through the forth time of companies in pulp and paper industry. Six companies were elevated to blue from black or red, while four companies were demoted. Companies that raised their rating have larger production capacities than companies than did not raise their rating. It is easy for large companies to invest in pollution control, but it is difficult for small companies to make investments to improve environmental management.

Table 6.3   The number of companies rated in PROPER PROKASIH

	May 1995	December 1995	October 1996	July 1997
Gold	0	0	0	0
Green	5	5	6	14
Blue	61	88	121	135
Red	115	115	80	116
Black	6	5	6	5
Total	187	213	213	270

Sources: Compiled from *Kompas*, 14 June 1995 and 30 December 1995; *Warta Ekonomi*, 11 November 1996 and BAPEDAL (1997a).

Table 6.4 Companies in pulp and paper rated in PROPER

Company	Province	October 1995	March 1997	Investment status, year operation started, production capacity
Aspex Pater, PT	West Jawa	Blue	Blue	F, 1985 Pa430,000t
Eureka Aba, PT	East Jawa	Red	Red	D, 1978 Pu30,000 Pa40,000t
Fajar Surya Wisesa, PT	West Jawa	Red	Blue	D, 1989 Pa500,000t
Indah Kiat Pulp & Paper Tnagerang, PT	West Jawa	Green	Green	F, 1979 Pa90,000t
Indah Kiat Pulp & Paper Riau, PT	Riau	Blue	Blue	F, 1989 Pu180,000t Pa650,000t
Indah Kiat Pulp & Paper Serang, PT	West Jawa	Red	Blue	F, 1991 Pa980,000t
Kertas Kraft Aceh, PT	Aceh	Red	Red	S, 1988 Pu170,000t Pa140,000t
Karya Tulada, PT	West Jawa	Blue	Stop	D, 1977 Pa9,000t
Kertas Bekasi, Teguh, PT	West Jawa	Red	Blue	D, 1976 Pu90,000t Pa150,000t
Kertas Padalarang, PT	West Jawa	Red	Red	S, 1923 Pu3,000t Pa8,000t
Kertas Basuki Rachmat, PT	East Jawa	Red	Red	D, 1971 Pu10,000t Pa13,000t
Kimsari Paper Indonesia, PT	N. Sumatra	Blue	Red	F, 1985 Pa4,000t
Onward Paper Utama, PT	West Jawa	Blue	Red	
Papyrus Sakti, PT	West Jawa	Black	Blue	D, 1976 Pa150,000t
Parisindo Pratama, PT	West Jawa	Blue	Blue	D, 1987 Pa20,000t

Pelita Cengkareng Paper, PT	West Jawa	Red	Black	D, 1976 Pa160,000t
Pakerin, PT	East Jawa	Blue	Blue	D, 1980 Pu150,000t Pa700,000t
Setia Kawan, CV	East Jawa	Red	Red	D, 1985 Pa30,000t
Suparma, PT	East Jawa	Blue	Blue	D, 1978 Pa150,000t
Surabaya Agung Industri Pulp & Kertas	East Jawa	Red	Blue	D, 1976 Pa340,000t
Surabaya Meka Box, PT	East Jawa	Red	Blue	D, 1973 Pa90,000t
Sari Morwa, PT	N. Sumatra	Black	Black	F, 1995 Pa4,000t
Tjiwi Kimia, PT	East Jawa	Blue	Red	D, 1978 Pa1,040,000t
Surya Zigzag	East Jawa		Green	D, 1989 Pa10,000t
Riau Andalan Pulp & Paper, PT	Riau		Blue	F, 1995 Pu750,000t Pa300,000t
Pindo Deli Pulp & Paper Mill, PT	West Jawa		Blue	F, 1978 Pa760,000t
Kertas Leces, PT	East Jawa		Blue	S, 1939 Pu130,000t Pa180,000t
Jaya Kertas, PT	East Jawa		Red	D, 1983 Pa200,000t
Gunung Jaya Agung, PT	West Jawa		Red	D, 1987 Pa36,000t

Note: F: Foreign investment; D: Domestic private investment; S: State enterprise; Pa: Paper; Pu: Pulp.

Sources: Compiled from *Warta Ekonomi*, 1 January 1996; BAPEDAL (1997b); and Indonesian Pulp & Paper Association (1999).

In the first ratings announcement, the five company names classified as green and the number of companies in each category were heavily featured by such mass media as Kompas, a leading newspaper in Indonesia. On the other hand, a six-month period for taking anti-pollution measures was granted to companies that received bad evaluations, and the names of those companies were not released in the first announcement. The names of companies rated as black were reported by the mass media following the ratings announcements in December 1995 and October 1996. Although the evaluation result of March 1997 was released in the form of a report in July, because the economic crisis started, the result was not widely reported in mass media.

Looking at 'black' companies, PT. Sari Morawa in North Sumatra was repeatedly classified as black (Table 6.5) from the second announcement through the fourth announcement. Except for PT. Sari Morawa, other companies classified as black in December 1995 had raised their rating by October 1996. However, five out of six companies classified as black in October 1996 were still classified as black in March 1997. It is difficult for the companies to improve their ratings within six months.

Although there were some limitations of the programme, more than a few companies improved their rating during the period from the first time in 1995 to the fourth time for 1997. It can be said that the programme was a success.

For what reason did companies make the decision to invest in anti-pollution measures? The information disclosure programmes on discharge of toxic substances in Western countries were models for PROPER PROKASIH. Generally, in Europe and the US, exposure of contamination leads to pressure from consumers, investors and surrounding residents, which forces companies to undertake anti-pollution measures. The World Bank (2001), which mentioned PROPER PROKASIH, offers the same explanation. Were

Table 6.5   Companies rated as black in PROPER PROKASIH (December 1995, October 1996, July 1997)

Name of company	Industry	Place	Year
PT. Raja Garuda Mas Panel Pekanbaru	Plywood	Riau	1995
PT. Tirta Mahakam	Plywood	East Kalimantan	1995
PT. Papyrus Sakti Paper Mill	Pulp and paper	Bandung	1995
PT. Sari Morawa	Paper	North Sumatra	1995, '96, '97
PT. Sico	Dye	North Sumatra	1995
PT. Madu Baru	Sugar	Jakarta	1996
PT. New Kalbar Processors	Rubber	West Kalimantan	1996, '97
PT. Kasin	Leather	East Kalimantan	1996, '97
PT. Sinar Obor	Leather	Jakarta	1996, '97
PT. Pelita Cengkareng Paper	Pulp and paper	West Jawa	1996, '97

Sources: Compiled from *Kompas*, 14 June 1995 and 30 December 1995; *Warta Ekonomi*, 11 November 1996; and BAPEDAL (1997b).

these kinds of effects observed in Indonesia? Afsah *et al.* (2000) sent a questionnaire survey to companies that were targeted in PROPER PROKASIH. They tried to identify the reason why companies undertook pollution control measures. It was shown that neither pressure from the stock market nor pressure from consumers was strong; however, pressure from the surrounding residents had an impact. As a result, managers became conscious of their factories' level of environmental measures, and they realized they needed to improve the environmental management of their factories.

It was not shown clearly in the above study, but it is possible that past cases might have an influence on the decision-making of these companies. For example, they may be influenced by the fact that the operation of their factories might be stopped, such as in the case of PT. Trebor, mentioned in the previous section, if it was found that the discharged pollutants exceeded the standard. Moreover, companies might face a boycott movement against their products if they pollute excessively. Entrepreneurs themselves understand the level of the countermeasures against pollution, and they consider it worthwhile to invest in pollution control to avoid the closing of their factories that may result from strict enforcement of regulations and to avoid boycott movements.

Some companies targeted by PROPER PROKASIH received low-interest loans for anti-pollution investment. Out of the eight companies in the paper and pulp industry that have received anti-pollution investment-oriented loans supported by Japan's OECF, six of them were companies targeted by PROPER PROKAISH. In some cases, when companies are considering anti-pollution investment, the availability of low-interest loans has determined whether a decision is made to actually invest.

PROPER

The Environmental Impact Management Agency was integrated into the Office of the State Minister for the Environment early in 2002, and the office changed its name from *Kantor Menteri* to *Kementrian*. This is regarded as an organizational promotion, but the office's status is not same as other ministries, which are called *departmen*. The internal organization was restructured and various programmes started. Preparation of PROPER, a new rating programme, was started in 2002.

The programme comprehensively evaluates environmental measurements, not only those related to water pollution but also those related to air pollution, hazardous waste management, environmental impact assessment and so forth.[12] Preparations for conducting ratings were begun in 2002, and performance was evaluated beginning in January 2003. The criteria of each evaluation category are as shown in Table 6.6, and these categories are modified from the categories of PROPER PROKAISH. Factories were evaluated based on 51 checkpoints, such as installation of a pollution control facility, emission levels, effluent levels and treatment of hazardous waste. The areas evaluated and the related environmental regulations are shown in Table 6.7.

Table 6.6 Rating in PROPER

Compliance status	Colour rating	Performance criteria
Compliant	Gold	Companies making efforts to achieve zero emission and reaching acceptable level.
	Green	Companies making efforts to prevent pollution and environmental destruction and reaching a level above the environmental standard.
	Blue	Companies making efforts to prevent pollution and environmental destruction and achieving the minimum environmental standard.
Non-compliant	Red	Companies making efforts to prevent pollution and environmental destruction but not achieving the environmental standard.
	Black	Companies making no effort to prevent pollution and environmental destruction.

Source: Compiled from Kementerian Lingkungan Hidup (2003), pp. 72–3.

Table 6.7 Area of evaluation in PROPER and related regulations

Area of evaluation	Related regulation
Wastewater pollution control	Governmental Decree No. 82 in 2001 Decree of the Minister No. 51 in 1995 Decree of the Minister No. 52 in 1995 Decree of the Minister No. 113 in 2003
Air pollution control	Governmental Decree No. 41 in 1999 Decree of the Minister No. 13 in 1995 Decree of the Minister No. 205 in 1996 Decree of the Minister No. 129 in 2003
Hazardous waste management	Government Decree No. 18 in 1999 Government Decree No. 85 in 1999 Decree of the Minister No. 68 in 1994 Decree of the Minister No. 01 in 1995 Decree of the Minister No. 02 in 1995 Decree of the Minister No. 03 in 1995 Decree of the Minister No. 04 in 1995 Decree of the Minister No. 05 in 1995
Environmental impact assessment	Governmental Decree No. 27 in 1999
Environmental management system	–
Resource conservation	–
Community relations and development	–

Source: Compiled from Ardiputra (2004).

The first rating for 85 companies was released in April 2004. No company was evaluated as gold. Eight companies (11 per cent) got a green rating. Blue amounted to 60 per cent, red was 25 per cent and black was 4 per cent.[13] The following companies were categorized as black: state-owned oil and gas firm PT. Pertamina UP IV in Central Jawa; textile producers PT. Kahatex II and PT. Prodomo, both in West Jawa; and paper producer PT. Papyrus Sakti Paper Mill in West Jawa. Incidentally, PT. Papyrus Sakti Paper Mill was listed as black in PROPER PROKASIH in 1995.

The second rating was released in December 2004. An attempt was made to expand the number of target companies, and the number of target companies in 2004 was 251, consisting of 117 companies in the manufacturing industry, including textiles as well as pulp and paper, 79 companies in the mining and energy industry, including coal and metal mining, and 53 companies in the agricultural and forestry, industry including sugar producers and palm oil producers. The evaluation rated 42 companies as black. The nine companies rated as green satisfied the requirements of environmental regulations. The 86 companies rated as red failed to obtain permits for their waste management. The 114 companies that satisfied the minimum environmental standards were rated as blue. Among the four companies listed as black in the first announcement, PT. Pertamina UP IV was raised to red, but the others remained in black.

Table 6.8 summarizes the ratings of companies listed in both the first and second announcement. Five companies increased their rating, while two companies stopped operation and seven companies decreased their rating. This suggests that the ratings of these seven companies were overestimated in the first announcement.

Table 6.9 shows the number of companies by rating and investment type. The ratings of foreign-owned firms are better than those of domestic private firms and state companies.

The third rating was released in August 2005. A total of 466 companies were rated. The results were as follows: 150 companies were given a red

Table 6.8 Changes in ratings from April 2004 to December 2004 in PROPER

		Second						
		Gold	**Green**	**Blue**	**Red**	**Black**	**Stop**	**Total**
First	Gold	0	0	0	0	0	0	0
	Green	0	6	2	0	0	0	8
	Blue	0	1	44	5	0	1	51
	Red	0	0	3	18	0	1	22
	Black		0	0	1	3	0	4
	Total	0	7	49	24	3	2	85

Sources: Compiled from Kementerian Lingkungan Hidup (2004) and Ardiputra (2004).

Table 6.9   Result of second announcement of PROPER

| | Rating | | | | | | Ratio of non-compliance |
	Gold	Green	Blue	Red	Black	Total	
Foreign	0	7	46	14	10	77	31.2 per cent
Domestic	0	2	35	38	23	98	62.2 per cent
State	0	0	33	34	9	76	56.6 per cent
Total	0	9	114	86	42	251	

Source: Compiled from Kementerian Lingkungan Hidup (2004).

Table 6.10   The number of companies rated in PROPER

	April 2004	December 2004	August 2005
Gold	0	0	0
Green	8	9	23
Blue	51	114	221
Red	22	86	150
Black	4	42	72
Total	85	251	466

Sources: Compiled from Kementerian Lingkungan Hidup (2004); Ardiputra (2004); and data on the website of the Ministry of Environment, http://www.menlh.go.id/proper/

rating, 72 were given a black rating, 221 companies got a blue rating and 23 companies got a green rating. No company got a gold rating.

The fluctuation in the number of companies in PROPER is shown in Table 6.10. Among four black companies listed in the first announcement, one company got a blue rating but one company remained black in August 2005. Fourteen companies got black ratings consecutively in December 2004 and August 2005. All of the announcements were taken up by mass media such as the *Jakarta Post*, *Kompas* and *Tempo*.

3.   Meaning and limitations of rating programmes

It is a fact that the rating programmes had a significant impact on pollution control in the target companies. As shown in PROPER PROKASIH, the number of companies that complied with the effluent standards increased significantly. The mass media widely featured the rating result, and this raised public awareness of the environment. It is also said that the rating programme shows the raison d'être of the Ministry for the Environment. The capabilities of the environmental administration have also been gradually increased from the end of the 1980s to the first half of the 1990s through implementing these programmes. Regulations concerning effluent and emission standards were created for every industry.

There are three major limitations on rating programmes.

First, the number of factories involved in the programmes is very limited. In 1997 there were 6,638 large-scale manufacturing companies and 15,748 medium-scale companies in Indonesia. The number of target companies in PROPER PROKASIH was only 270 at the final rating in 1997. In PROPER, the number of target industries was expanded to include mining and plantations. However, the number of the companies was still 466 in the third ratings announcement in August 2005. Although human resources in the Ministry of the Environment concentrate on collection, verification and evaluation of the data for ratings, the coverage is still limited. It is very difficult to obtain a sufficient budget and adequate human resources for rating all manufacturing factories in Indonesia.

The second limitation is the low number of indicators employed to determine the environmental burden. In PROKASIH and PROPER PROKASIH, from among the various existing pollution indexes, concern is concentrated on BOD or chemical oxygen demand (COD). The criterion used by PROKASIH to evaluate company performance is the change in the BOD load.[14] The companies classified as black, the worst level, were the companies in industries where the BOD load is generally higher than other industries, such as pulp and paper, as well as dyeing. However, one problem is that indexes of pH levels and heavy metals, etc., are not considered to be as important as BOD or COD. According to the Environmental Impact Management Department in Bandung of West Jawa, 108 companies should have had pollution control facilities in the city in 2002. Table 6.11 shows the state of installation of pollution control facilities. Among 18 electroplating factories that were not targeted in PROKASIH and PROPER PROKASIH, only two companies installed effluent treatment facilities. Among 67 textile companies that were the major target in PROKASIH and PROPER PROKASIH, 61 companies installed effluent treatment facilities. Which index these programmes emphasize affects the impact of the pollution control measures in each industry. In addition, seven out of 35 companies for which the Environmental Management Department of West Jawa conducted wastewater examinations in August 2000 were not even able to meet the standard for pH, which is a comparatively easy-to-attain standard. The pH value at three companies exceeded ten again when the investigation was conducted in October, which indicated very high alkali.

In PROPER, the number of indicators employed to evaluate the performance of companies increased but is still too limited. WALHI, an environmental NGO, criticized the programme, saying that many companies involved in cases of illegal logging have not been given poor ratings.[15]

The third limitation is weakness of the threat to stop the operation of companies that pollute. Due to the weakness of the threat, it is possible for companies to completely ignore ratings. Several companies were repeatedly placed in the black and red categories in PROPER PROKASIH and PROPER. Announcing a company name is weaker than stopping operation of a factory, in the sense of legal force, but the possibility of having their operations

Table 6.11   State of installation of waste treatment facilities by companies in Bandung

Industry	Number of companies	Waste treatment facility			No treatment facility
		Chemical treatment	Biological treatment	Incineration	
Textile	67	55	6		6
Medicines	3	2	1		
Electroplating	18	2			16
Paper	1	1			
Paint	1				1
Soap	1	1			
Cable	1			1	
Leather	4	4			
Cooking oil	2	2			
Food	10	1			9
Total	108	68	7	1	32

Source: Data provided by the Environmental Impact Management Department in Bandung City, in March 2002.

stopped is a threat to the companies listed in the rating programme. Stricter enforcement of the law would stimulate a better response from owners of companies categorized as black and red. Efforts should be made to promote the strengthening of enforcement capabilities and the shortening of the bureaucratic process for stopping operations of polluters.

Ratings are useful in the process of developing the enforcement of environmental regulations, when the Ministry of the Environment does not have enough authority to stop operation of a polluting factory. The Ministry of the Environment should not rely solely on rating programmes. Rating should be used as a method to raise environmental awareness and to supplement weak enforcement.

Conclusion

The development of a pollution control policy consists of three parts: institutional development for protecting the environment, regulation to control industrial activity that discharges pollutants and enforcement of the regulations. The rating programme in Indonesia has entered the period of strengthening enforcement.

The success of PROPER PROKASIH was renowned not only domestically but also internationally. Although PROPER PROKASIH is recognized as the prototype of similar programmes in developing countries, the first rating in PROKASIH was conducted in Jakarta in 1991. This rating attracted mass media attention and raised environmental awareness not only of the public and but also of managers of companies. The Ministry of the Environment defined its role in environmental management through these rating programmes.

However, as stated above, these programmes have limitations. The most significant limitation is the small number of target companies. Although there are many factories located in Indonesia, less than 2 per cent of large- and medium-scale factories are covered by these programmes.

Second, the ratings are not powerful enough to force companies to invest in pollution control. Some companies have shown up repeatedly in the black category. The threat to stop factory operation due to pollution could make the rating programme more effective. The combination of the rating programme and the penalties for non-compliant companies should be emphasized more. For this, capabilities need to be further developed so that adequate evidence concerning pollution can be collected. In addition, the procedure for stopping factories' operation should be simplified.

Where there are many violators of the pollution control standard, a rating programme is not an alternative to a command and control policy. The priority of environmental policy should be placed on the enforcement of effluent and emission standards. It is not realistic to increase the budget and human resources for a rating programme to cover all factories in Indonesia. It is better to reduce the amount of resources used to evaluate good companies and to increase effort spent on monitoring companies in violation of effluent or emission standards.

However, a rating programme can attract mass media attention, thereby contributing to raising public awareness and building the capacity to handle environmental issues in central and local governments. A rating programme can be used as a supplement to command and control policy or other pollution control policies during the period when developing countries are strengthening the enforcement of their environmental policies.

Notes

1. The official name of PROPER PROKASIH is Program Penilaian Kinerja Perusahaan/ Kegiatan Usaha (Corporate Activity Evaluation Program), and PROKASIH is the abbreviated name of Programme Kali Bersih (Clean River Programme).
2. World Bank (2000) said that India, Bangladesh, Thailand and Venezuela were also considering a similar programme.
3. Cribb (1990, 1998) summarized the topic in detail.
4. In Bandung in 1992, the level of the bribe was 20,000 rupiahs for one person. On the other hand, the staff of the provincial industrial pollution management team of West Jawa (TKP2D) did not receive a bribe. It is reported that special advice was given from the staff of TKP2D.
5. Based on Braadbaart (1995).
6. Refer to the Office of the State Minister for Population and the Environment (1990).
7. Refer to *Jakarta Post*, 30 December 1991.

8. BOD pollution loads are based on an internal document of the Environmental Impact Management Agency.
9. Based on Aditjondro (1991).
10. See PROPER-PROKASIH Team and PRDEI (1995).
11. The statistics which were released in 1996 and published by the newspapers, etc., are inconsistent with the statistics of the World Bank (1997).
12. Based on the interview in March 2002 with the deputy assistant minister on infrastructure and manufacturing industry of the Ministry of the Environment and on a press briefing on PROPER in October 2004, which was published in December 2004.
13. 73 pages of Kementrian Lingkungan Hidup (2003).
14. This point is based not only on Section 2 and Section 3 but also on the data of BAPEDAL (1994, 1995a, 1995b). BAPEDAL, (1995a and 1995b, are the FY1995 activity reports of PROKASIH in Jakarta and West Jawa, respectively. No contaminants other than BOD and COD are mentioned in 1995a and 1995b).
15. 'Black Stamps for Polluters' *Tempo*, 5 September 2005.

References

Aditjondro, George (1991) 'Boycott Ethics', *Environesia*, 2, 3–5.

Afsah, Shakeb, Allen Blackman and Damayanti Ratunanda (2000) 'How Do Public Disclosure Pollution Control Programs Work? Evidence from Indonesia', Discussion Paper 00–44, Washington, DC, Resources for the Future.

Afsah, Shakeb, Benoît Laplante and Nabiel Makarim (1996) 'Program-Based Pollution Control Management: The Indonesian PROKASIH Program', Policy Research Working Paper, No. 1602 Washington, DC, World Bank.

Ardiputra, Isa Karmisa (2004) *Proper: Program Penilaian Peringkat Kinerja Perusahaan*, Jakarta.

Badan Pusat Statistik (2001a) *Large and Medium Manufacturing Statistics, 1999*, Jakarta.

Badan Pusat Statistik (2001b) *Environmental Statistics of Indonesia 2000*, Jakarta.

BAPEDAL (1994) 'PROKASIH: Evaluating the Last Four Years and Looking to the Future', Jakarta.

BAPEDAL (1995a) *Laporan Ringkasu: Pelaksanaan Program Kali Bersih (PROKASIH) di Propinsi Dki. Jakarta.*

BAPEDAL (1995b) *Laporan Ringkasu: Pelaksanaan Program Kali Bersih (PROKASIH) di Propinsi Jawa Barat*, Jakarta.

BAPEDAL (1996) *Cleaner Production Action Plans*, Jakarta.

BAPEDAL (1997a) 'Summary of the March 1997 PROPER PROKASIH Evaluation: July 1997', press release.

BAPEDAL (1997b) *Hasil Penilaian PROPER PROKASIH bulan Juli 1997*, Jakarta.

Braadbaart, Okke (1995) 'Regulatory Strategies and Rational Polluters: Industrial Wastewater Control in Indonesia, 1982–1992', *Third World Planning Review*, 17(4), 439–58.

Cribb, Robert (1990) *The Politics of Environmental Protection in Indonesia*, Monash University, Australia, The Centre of Southeast Asian Studies.

Cribb, Robert (1998) 'Environmental Policy and Politics in Indonesia', in Uday Desai (ed.), *Ecological Policy and Politics in Developing Countries: Economic Growth, Democracy, and Environment*, Albany, State University of New York Press.

Dana Mitra Lingkungan (2000) *A Final Report on US-AEP and DML Technical Cooperation for the Improvement of the Small & Medium Industry's Performance to Recover from the Crisis*, Jakarta.

El Khobar Muhaemin Nazech (2001) 'Study on Indonesia Industrial Sectors' Contribution to Sustainable Development: Draft Final Report', United Nations Industrial Organization.

Ex Corporation (1992) *Indonesia: Resources and the Environment – Toward an International Cooperation for the Environmental Management*, Tokyo.

Hardjasoemantri, Koesnadi (1994) *Environmental Legislation in Indonesia*, Yogyakarta, Gadjah Mada University Press.

Kementerian Lingkungan Hidup (2003) *Status Lingkungan Hidup Indonesia 2002*, Jakarta.

Kementerian Lingkungan Hidup (2004) *Bahan Press Breifing: Program Penilaian Peringkat Kinerja Perusahaan (PROPER) Oktober 2004*, Jakarta.

Kojima, Michikazu and Norio Mihira (1995) 'Economic Development and Pollution in Indonesia', in Kojima, R., Y. Nomura, S. Fujisaku and N. Sakumoto (eds), *Development and the Environment – The Experiences of Japan and Industrializing Asia*, Tokyo, Institute of Developing Economies, 355–74.

Inoue, Makoto and Michikazu Kojima (2000) 'Indonesia', in Japan Environmental Council (ed.), *The State of the Environment in Asia 1999/2000*, Tokyo, Springer, 83–95.

Indonesia Pulp & Paper Association (1999) *Directory 1999*, Jakarta.

López, Jorge Gracía, Thomas Sterner and Shakeb Afsah (2004) *Public Disclosure of Industrial Pollution: The PROPER Approach for Indonesia?*, Washington, DC, Resources for the Future.

MacAndrews, Colin (1994) 'The Indonesian Environmental Impact Management Agency (BAPEDAL): Its Role, Development and Future', *Bulletin of Indonesian Economic Studies*, 30(1), 85–103.

Office of the State Minister for Population and the Environment (1990) *Report on the International Workshop on PROKASIH (Clean River Programme)*, Jakarta.

Pargal, Sheoli and David Wheeler (1996) 'Informal Regulation of Industrial Pollution in Developing Countries: Evidence from Indonesia', *Journal of Political Economy*, 104(6).

Plansearch Associates (1999) *The Market Potential for Pollution Control Equipment in Indonesia, Volume II and III, A Report for United States-Asia Environmental Partnership Program*.

PROPER-PROKASIH Team and PRDEI (1995) *What is PROPER?: Reputational Incentives for Pollution Control in Indonesia*, Jakarta.

Rock, Michael T. (1999) 'Initiating Environmental Behavior in Manufacturing Plants in Indonesia', *Journal of Environment and Development*, 8(4).

Rock, Michael T. (2002) *Pollution Control in East Asia: Lessons from Newly Industrializing Economies*, Washington, DC, Resources for the Future.

Santosa, Happy (2000) 'Environmental Management in Surabaya with Reference to National Agenda 21 and the Social Safety Net Programme', *Environment and Urbanization*, 12(2), 175–84.

State Minister for Environment and United Nations Development Program (1997) 'Agenda 21 – Indonesia: A National Strategy for Sustainable Development', Jakarta.

Villamere, John and Anhar Kusnaedi (1992) *Environmental Assessment Training for INKINDO: Final Report*, EMDI Project, Jakarta.

Vincent, Jeffrey R., Jean Aden, Giovanna Dore, Magda Adriani, Vivianti Rambi and Thomas Walton (2002) 'Public Environmental Expenditures in Indonesia', *Bulletin of Indonesian Economic Studies*, 38(1), 61–74.

World Bank (1997) 'Creating Incentives to Control Pollution', *DECnotes*, No. 31.

World Bank (2000) *Greening Industry: New Roles for Communities, Markets and Governments*, Washington, DC, Oxford/World Bank.

World Bank (2001) *Indonesia: Environment and Natural Resource Management in a Time of Transition*, Washington, DC, World Bank.

7
Industrial Pollution Control in India: Public Interest Litigation Re-examined

Yuko Tsujita

Introduction

The salient feature of India's pollution control is the active role played by the judiciary. There are an increasing number of public interest litigation (PIL) cases and other legal efforts, which are playing an important role in environmental policy implementation and legal compliance. PIL, in addition to formal lawsuit procedures, allows the general public to invoke the warrant jurisdiction of the court, sometimes even by sending a postcard directly to the Supreme Court or High Court. Buses, taxis and three-wheeled vehicles used for public transportation in the capital city are now fuelled with compressed natural gas (CNG). This is one example where PIL was effective in ameliorating the air pollution problem by leading to a Supreme Court order that public transportation must use CNG. There are now so many environmental lawsuits that every Friday is 'Green Friday', when Supreme Courts deal with environment-related cases (Agarwal *et al.*, 1999).

Judicial activism plays a significant role in awakening the general public's concerns about environmental issues and causing the administration to initiate policy implementation and legal compliance. At the same time, there are suggestions that could improve environmental court orders, particularly their technical rationality. Efforts at improving orders, however, will not necessarily result in solutions for environmental and pollution problems. In the case of industrial pollution, in order to pass the periodic pollution monitoring inspections, some firms operate emission treatment plants only during the inspections or pay bribes to the inspection officials (Lindstad, 1999). This means that industrial pollution control may be enhanced by some mechanisms other than administrative regulation or judicial initiatives.

Do firms take measures to meet environmental standards simply because of government regulation? One study that suggests that the market and community play large roles is the World Bank (2000), which pays attention to the market and community as informal regulators. It shows that information on firms' pollution control does affect the price of stock. In India, one of the leading environmental NGOs, Centre for Science and Environment, has graded the environmental performance of large companies in a few hazardous industries. So far, few studies in India have empirically examined the influence of the rating programme on the stock market,[1] and the role of the stock market in developing countries is still questionable.

Local communities have also been regarded as important informal regulators in a few previous studies. The World Bank (2000) showed that units located in communities where local residents' educational and income levels are higher are more likely to take emission measures. By the same token, Murthy and Prasad (1999) showed that among 100 units in 13 states and 11 different industries in India, units located in the areas where residents' voting rate is higher are more likely to comply with the Bio-Chemical Oxygen Demand (BOD) standard. That means that the local community does affect the emission control of units. On the other hand, Pargal *et al.* (1997) rejected the hypothesis that firms located in areas of higher educational and income levels are more likely to practice emission control, although quite a number of firms cited complaints from local communities and NGOs as a reason to take emission control. The reason why the hypothesis is not statistically significant is assumed to be PIL. The environmental PIL has been evaluated as a valuable tool, similar to an 'indirect market-based tool' (Sawhney, 2003). That is why the improvement of court orders is often pointed out as a means of improving pollution control. However, if you look at court cases carefully, the orders are not always implemented, although the orders are crucial up to a certain point in pollution control. Therefore, I will examine advantages and disadvantages of judicial activism, especially from the viewpoint of community participation, in order to show that judicial activism is not a permanent solution to environmental problems. Accumulation of specific examples is necessary to prove that communities' monitoring of pollution is more effective than continuing the status quo, which would be to further enhance the judicial activism.

In Section 1, after briefly explaining the administrative and legal framework of industrial pollution control, I will discuss why the judiciary has to be active. In Section 2, I will explain three Supreme Court cases. In Section 3, I will examine advantages and disadvantages of judicial activism and analyse the role of the community under judicial activism.

1.   Industrial pollution control: laws, regulations and enforcement

In addition to the effects of the early development of heavy industry under the import-substitution industrial policy in force since independence in 1947, water pollution was exacerbated by the climate, with its low rainfall, and by geographical conditions such as rivers with low gradients, even during the low economic growth period between the 1950s and 1970s. Since the start of far-reaching economic reforms in 1991, competition among states to attract

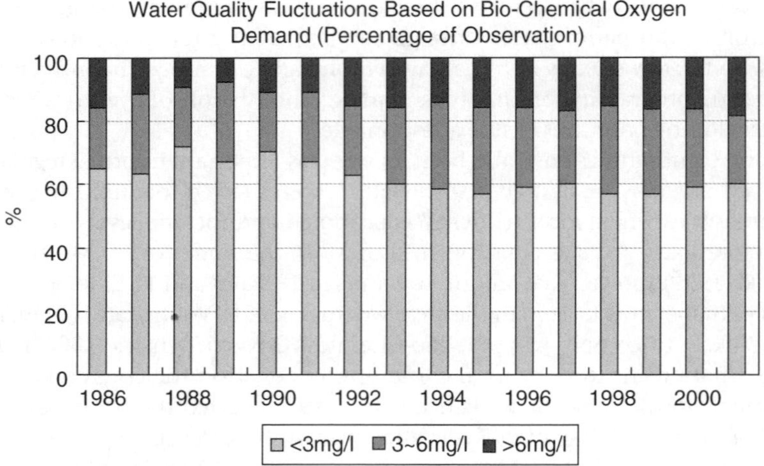

Figure 7.1   The share of BOD in national monitoring points
Source: Ministry of Environment and Forests, 2003.

investment has grown increasingly severe, enabling companies that emit pollutants to invest in India (Kuik *et al.*, 1997). Figure 7.1 shows fluctuations in the
water quality in terms of BOD under the water quality monitoring programme.
In the late 1980s, more than 60 per cent of the monitoring points had less than
3 mg/l; however, the percentage decreased over the 1990s. Water bodies may
have become more contaminated recently, due to water scarcity, low rainfall
and the increasing pollution load (Ministry of Environment and Forests, 2003).

By now, India's legal and administrative framework for pollution control has
become well developed. At the same time, the implementation of environmental policies and legal compliance are still questionable. That is why the judiciary
has taken so much initiative in environmental and pollution issues. Let us first
survey the overall picture of laws and regulations and their enforcement.

Strict laws and regulations

The first event to spark environmental legislation after independence was the
1972 United Nations Conference on the Human Environment in Stockholm,
which India followed with the 1974 Water Prevention and Control of Pollution
Act and the 1981 Air Prevention and Control of Pollution Act. These were, however, more end-of-pipe laws than attempts to eliminate the causes of pollution.

The next event that triggered environmental action was the Bhopal gas
tragedy in 1984, which served as a wake-up call pointing to the need to have
comprehensive environmental laws that include environmental impact assessments (EIA), and led to creation of the Environmental Protection Act and the
Ministry of Environment and Forests in 1986.[2] Through the Environmental

Table 7.1   Main environmental laws

Environment and industry	Environment (Protection) Act, 1986
	Boilers Act, 1923
Air pollution	Air (Prevention and Control of Pollution) Act, 1981
	Motor Vehicles Act, 1989
Water pollution	Water (Prevention and Control of Pollution) Act, 1974
	Water (Prevention and Control of Pollution) Cess Act of 1977
Radiation	Atomic Energy Act, 1982
Mining	Mines and Minerals (Regulation and Development) Act, 1957
	Coal Mines (Conservation and Development) Act, 1974
Insecticides	Poison Act, 1919
	Insecticides Act, 1968
Land	Urban Land (Ceiling and Regulation) Act, 1976
	Coal Mines (Conservation and Development) Act, 1974
Factories	Factory Act, 1948
	Industries (Development and Regulation) Act, 1951
Forests	Indian Forest Act, 1927
	Wild Life (Protection) Act, 1972
	Forest (Conservation) Act, 1980
Pollution compensation	Public Liability Insurance Act, 1991
	National Environment Tribunal Act, 1995
	National Environment Appellate Authority Act, 1997
Energy conservation	Energy Conservation Act, 2001

Sources: Central Pollution Control Board, 2001; and Nomura and Endo, 1994.

Protection Act 1986, EIAs, environmental audit reports, environmental clearances for specific projects and public hearing as a part of EIAs have been introduced gradually. Moreover, India's first regulations on hazardous waste management were enacted based on the Environmental Protection Act, although action in this area is behind that for air and water pollution. Later in 1991, India enacted the Public Liability Insurance Act for helping victims of hazardous chemical accidents. This act establishes a compulsory liability insurance system meant to quickly provide victims with the minimum basic compensation, but victims can also file lawsuits to demand more than basic compensation. This law is considered to be one of the world's three major pollution relief institutions, along with Japan's Pollution-related Health Damage Compensation Programme and the United States' Superfund Law (Nomura and Endo, 1994).

India has a comparatively well-developed system of environmental laws to deal with pollution (Table 7.1). Each state government can enact laws on matters such as water, which the constitution puts under the states' purview, as well as laws and regulations stricter than those of the union government.

Additionally, state pollution control boards can order closures of units that do not comply with air and water emission standards. However, officials in charge on site have considerable discretion in these matters.

The Rio Summit in 1992 led to the enactment of environmental policies and prioritization of environmental issues in India. Needless to say, the government has been aware of the need for environmental considerations in its new industrial policy even after beginning economic reforms in 1991. For the summit in 1992, the government issued a Policy Statement for Abatement of Pollution. This statement announced strengthening of regulations and introduction of economic incentives from a long-term view of environmental problem-solving. It is probably noteworthy that the statement mentioned the necessity of shifting from end-of-pipe treatments to development of cleaner technology, among other matters. However, as it does not specifically name the responsible administrative agencies, it is not clear how this statement fits into the overall scheme. Soon after that statement, the National Conservation Strategy and Policy Statement on Environment and Development was announced. This is drawn up within the framework of the 1992 policy statement, i.e., it includes technology development, pollution controls at the source, 'polluters pay' principle, conservation of protected areas and public participation in the decision-making process. Furthermore, the Environmental Action Plan in December 1993 mentions pollution control and reduction of wastes, especially hazardous waste, as one of the seven highest-priority environmental problems that India should tackle.[3]

In sum, command and control regulations have been the main type of measure used to control industrial pollution, although other measures such as economic incentives and voluntary agreements between government and 17 hazardous industries have also been introduced.[4] Despite this solid framework of regulations, the enforcement of regulations is still far behind. Next, I will discuss the four main reasons why the enforcement of pollution regulations and laws is weak.

Background of weak enforcement

Low priority in government policy

Industrial pollution control has been given a low priority under the government's environmental policy. It is apparent that forestry has been given higher priority among the environmental issues. Table 7.2 shows that the budget allocation for forestry is much higher than those of other environment issues (i.e., environment and national river protection). During the 1980s, and even during the 1990s when the budgets of other environmental issues were higher than forestry, actual expenditures turned out to be lower than those of forestry. At the state government level, this trend is much clearer (Table 7.2).

Furthermore, out of the external aid received between 1995 and 2000 ($98.87 billion), the largest components were allocated to infrastructure

Table 7.2 Environmental plan outlays and expenditures of central and local governments

		7th Plan (85/86–89/90)		8th Plan (92/93–96/97)		9th Plan (97/98–01/02)
		Outlay	Expenditure	Outlay	Expenditure	Outlay
Central Government	Environment	110	129.38	325	406.94	859.84
	NRCD	240	192.15	350	314.06	700
	Forests and wildlife	155	100.92	250	422.26	854
	NAEB	292	256.85	275	488.67	600
	Total	797	679.3	1200	1631.93	3013.84
States and Union	Environment	7,791	6,021	15,311	13,093	269,852
Territories	Forestry	141,239	163,077	355,687	360,278	630,016

Note: NRCD stands for National River Conservation Directorate, and NAEB stands for National Afforestation and Eco-Development Board.
National Watershed Development Board.
Under the 8th plan, no data on Union Territories is available.
Source: Planning Commission, 2001.

in sanitation and hygiene ($22.21 billion), followed by biodiversity ($15.13 billion), watershed development ($14.26 billion), forestry ($796 million) and land ($794 million) (Confederation of Indian Industry, 2002). Obviously, the priority area is infrastructure and land-related areas. Industrial pollution, on the other hand, was allotted only $670 million, which is only 6.6 per cent of all environmental aid.

The reason why provision of infrastructure is emphasized so much is mainly that environmental issues have been regarded as poverty and infrastructure development issues. In the 1972 United Nations Conference on the Human Environment in Stockholm, then Prime Minister Indira Gandhi made a speech as follows. 'The environment cannot be improved in conditions of poverty. Nor can poverty be eradicated without the use of science and technology' (quoted from Divan and Rosencranz, 2001, p. 32). In the same year, the government announced the Minimum Needs Programme, which included drinking water, among other matters, as a priority area and, furthermore, in the 1980s, when the government's prescription for poverty alleviation called Revised 20 Programmes was announced, drinking water in rural areas, social forestry and non-conventional energy were included. The government prioritized the provision of infrastructure, such as drinking water, and tried to alleviate poverty by providing infrastructure.

After the Bhopal incident, the Seventh Five-year Plan (1985/6–1989/90) regarded the negative aspect of development as one of the causes of environmental degradation, which represented a dramatic change from the conventional view in the government documentation. However, environmental issues are still linked to underdevelopment, and this viewpoint was repeated in different environmental statements in the 1990s. Due to lack of environment-related infrastructure, such as drinking water and sanitation, India has to tackle both the issue of lack of infrastructure and the issue of pollution control. Consequently, infrastructure development is regarded as being much more important and urgent than controlling pollution.

Limited resources and limited capacity of state pollution control boards

Each state government has a pollution control board (SPCB) that is in charge of issuing environmental authorization for establishment and expansion of units and environmental authorization for units' operation standards for air and water pollution, developing environment-friendly technology, controlling pollution through inspection of industrial units, regulating the location of industries, collecting and disposing of hazardous waste and disseminating information on the prevention and control of pollution. The government has published a number of the SPCBs' evaluation reports in which it is disclosed that there is inappropriate control of water and air pollution due to lack of resources in personnel, finance and infrastructure.

The National Planning Commission (Planning Commission, 2000) evaluated SPCBs during the Eighth Five-year Plan (1992/3–97/8). Some of the important issues, in terms of the SPCBs' organization, finance and performance, are highlighted in this report. The issues include the relatively low number of technical staff at SPCBs in industrialized states, the fact that the number of units registered with SPCB is lower than the number shown in industrial statistics, the availability of only nine hazardous waste disposal sites nationwide, insufficient empowerment of SPCBs and so on. As a result of these problems, the percentage of units that install pollution equipment is small and, what is worse, the percentage of units which comply with the environmental standards is even smaller (Table 7.3). Obviously, establishment of regulations for installing facilities to control pollution is not sufficient to induce units to satisfy the prescribed standards for air and water, given the limited resources and limited capacity of the state pollution control boards.

Small-scale industries' slow action for pollution control

Small-scale industry (SSI) has been promoted by the government since independence, mainly in order to boost general mass employment, promote industrialization in the backward areas and redress regional inequalities and income distribution.[5] Currently, SSI accounts for approximately 40 per cent of total industrial production and approximately 35 per cent of export. However, 90 per cent of the units in the SSI category do not take any pollution measures (*Down to Earth*, 15 October 2002). Limited financial resources, unavailability of technical support and lack of space to install pollution equipment are the main reasons behind this.

The number of monitorings and inspections of SSI units by state government is set to be less than those of larger units. Unless SSIs are operated in an industrial estate, the cost of monitoring them is high since these units are often non-registered and located in backward areas or isolated places. Due to these reasons, it is not easy for SPCBs, with their limited budget and personnel resources, to inspect and monitor SSIs fully, even though it is estimated that 40 per cent of water pollution is discharged from SSIs (CPCB, 2001).

Few NGOs involved in industrial pollution

Recently, there has been an increasing number of government-NGO partnership projects in the field of environmental and pollution control. The Ministry of Environment and Forests, for example, set up an NGO cell in 1992, while the Central Pollution Control Board (CPCB) also established an NGO cell in 1994 to be in charge of providing training programmes for NGOs. The Ninth Five-year Plan of government (1997/8–2001/2) emphasized that NGOs, farmers and communities are expected to play an active role in solving environmental problems.

Table 7.3 Industrial units according to their pollution and control status

State	High pollution units (HPUs)				Water pollution			Air pollution		
	Number of HPUs inventorized	Estimated number of HPUs	Number inventorized as % of number estimated	Number of HPUs with facilities to satisfy standards	No. of water polluting units	% of units with effluent treatment plant	% of units satisfying standards	No. of air polluting units	% of units with effluent treatment plant	% of units satisfying standards
Andhra Pradesh	220	550	40.00	96	2,820	90.85	90.85	2,520	79.84	79.84
Assam	15	33	45.45	60	95	30.52	13.68	86	38.57	32.56
Bihar	62	226	27.43	83	116	70.69	29.31	1,386	40.55	40.55
Gujarat	200	551	36.30	95	8,098	52.72	32.16	5,757	59.74	54.87
Haryana	230	203	113.30	40	2,580	63.49	53.72	1,513	74.88	26.76
Karnataka	120	273	43.96	91	8,015	59.50	57.83	6,902	59.79	46.33
Kerala	24	78	30.77	92	2,250	51.95	35.60	1,528	62.04	24.41
Madhya Pradesh	103	371	27.76	99	526	78.90	–	526	68.63	68.63
Maharashtra	335	845	39.64	96	7,169	82.29	62.29	7,008	72.60	58.86
Orissa	92	111	82.88	–	–	–	–	–	–	–
Punjab	58	413	14.04	76	3,280	49.72	49.72	8,299	17.62	17.62
Rajasthan	49	347	14.12	98	692	80.60	–	430	91.00	–
Tamil Nadu	188	1,280	14.69	98	6,338	41.23	–	6,998	86.12	–
Uttar Pradesh	735	1,438	51.11	84	454	81.94	48.90	281	90.75	80.07
West Bengal	73	400	18.25	81	62	96.77	59.68	6,188	–	–

Note: Industrial units that fall in the specific 17 categories are highly polluting units.
Source: Planning Commission, 2000.

Table 7.4   Activities of environmental NGOs

Activity areas	No. of NGOs
Forestry	862
Non-conventional energy	464
Water supply, sanitation and solid waste management	249
Water and soil pollution	236
Wasteland development	201
Agriculture, organic farming or food production	135
Wildlife	106
Ecosystem, conservation areas	83
Watershed management	64
Biodiversity	52
Industrial and/or mining pollution	43
Dam construction and resettlement	34
Natural calamity (flood, drought etc)	25
Air pollution	12
Hydro/thermal/nuclear plants	8
Eco tourism	4
Noise	1
Total No. of NGOs in the list	1478

Note: Most NGOs are engaged in more than two activities, therefore the cumulative number of
NGOs outnumbers the total number of NGOs in the list.
Source: Compiled from WWF India, 1999.

WWF India (1999) lists 1,478 environmental NGOs in India as of 1997. This list does not necessarily cover all the environmental NGOs; however, the general picture can be grasped from an analysis of this list. Two features should be mentioned. First, 42.8 per cent of NGOs were established in the 1980s, which exceeds the 15.4 per cent established in the 1970s and the 24.3 per cent in the 1990s. That means the establishment of environmental NGOs and judicial activism emerged at the same time, in the 1980s. Second, the number of NGOs in industrial pollution is very limited. Table 7.4 shows that 58.3 per cent of environmental NGOs are engaged in forestry, followed by 31.4 per cent in energy and non-conventional energy, 16.8 per cent in water supply, sanitation and solid waste, 16 per cent in water and soil pollution and 13.6 per cent in wasteland development, and all of these have quite a few central government funding schemes. There are only 43 NGOs in industrial pollution and mining, and 35 of these are engaged in industrial pollution control. Most of these NGOs work in limited geographical areas, and therefore NGOs are not capable of exerting pressure for pollution control all over the country.

Judicial activism and civil society

As has been discussed, the low priority of pollution control in government policy, administrative problems, SSIs' sluggishness in pollution control and the low

Table 7.5   Number of environmental PIL

	Supreme court		Andhra Pradesh High court	
	No. of filed cases	No. of disposed cases	No. of filed cases	No. of disposed cases
1990	6	4	2	2
1991	7	7	3	3
1992	5	4	3	1
1993	1	1	3	3
1994	7	3	7	6
1995	6	4	10	9
1996	9	6	29	27
1997	6	5	74	61
1998	6	0	52	33
1999	5	2	76	51

Source: Prasad, 2004.

number of NGOs involved in industrial pollution has led to the government's lax enforcement of environmental laws and regulations. In contrast, the judiciary has made active interventions in this area. PIL, as mentioned in the Introduction, is a popular tool for brining attention to environmental and pollution issues. Since the 1980s, the number of environmental court cases has increased dramatically. Sato (2001) categorized interim orders of PILs from 1980–96 and found that 39 per cent of the PILs involved public policy, including environmental issues, and this exceeds the figures for other categories, such as police and jail (21 per cent), administration (18 per cent), women and children (8 per cent), labour (7 per cent) and urban issues (7 per cent). He further mentioned that, in the early 1990s, 15,000 letters per year were sent to the Supreme Court to apply for PILs. Only a few of them have been registered as PILs, or in some cases transferred to High Court, and therefore the number of environmental PILs filed and disposed is only five to nine cases per year, a figure which did not increase over the 1990s (Table 7.5). On the other hand, High Courts of several states have been persuaded to set apart a greater share of judicial resources for environmental cases, such as establishing the 'green bench' (Divan and Rosencranz, 2001). In Andhra Pradesh High Court, for example, the number of environmental PILs did increase in the 1990s (Table 7.5). Prasad (2004) further showed that court orders are often favourable to complainants, i.e. 38.9 per cent of the orders in the Supreme Court and 61.2 per cent of the orders in Andhra Pradesh High Court in the 1990s were favourable to complainants.

Through PIL petitions, the judiciary can order the government and/or firms to comply with pollution standards. The strict judicial orders have frequently been reported by the media, which no doubt has led to a boost in the general

Table 7.6   Petitioners of industrial pollution orders in the supreme court

Petitioner	No. of Cases
Administration	11
Firms	9
Lawyers	14
Individuals	8
Organizations (incl. NGOs)	11
Total	53

Note: An Internet search for 'industrial pollution' gets 72 hits (as of February 1, 2003) and 53 of those are related to industrial pollution, according to the content of order.
Source: Compiled from the Supreme Court of India's Judgment Information System (http://www.indiancourts.nic.in/).

public's concern over environmental and pollution issues. Triggered by the court orders, the administration also has taken initiatives in some of these issues. Therefore, the role of the judiciary should be highly appreciated.

At the same time, judicial activism hinders monitoring of pollution emissions by local residents because community participation in inspection and monitoring does not seem to be promoted under judicial activism. The public has the right to take legal action against units that discharge pollution and to complain to both central and state governments.[6] However, these means of redress do not seem to be fully utilized by the public. At most, filing a case through PIL could be a major means utilized when the public wishes to raise concerns over environmental issues. However, PIL is used more, actually, by environmental specialists or those who are particularly interested in environmental or pollution issues. The list of petitioners in the case of industrial pollution in the Supreme Court (Table 7.6) shows that there are 14 lawyers and eight individuals who filed cases, although there are also 11 NGOs and organizations. Having investigated each case on the list, lawyers and individual petitioners are mostly not local residents of polluted areas because PIL allows any concerned citizen to file a case on behalf of the larger community that is suffering from pollution or a member of the public who has generally suffered. For example, M. C. Mehta, one of the most prominent Supreme Court environmental advocates, filed a number of environmental cases in different parts of the country, although he does not necessarily suffer from pollution or environmental degradation in each case.

No doubt judicial activism is effective until a court order is issued; however, once orders are announced, then the responsibility for implementation is handed over to the administration and/or industrial units. That is why the implementation of court orders is frequently delayed or insufficient. As discussed, SPCBs are not always active in monitoring and inspection mainly due to lack of resources. Therefore, the role of community as an informal regulator or stimulator of local administration is crucial in pollution control. Below, we will look at three Supreme Court cases on pollution control in the leather industry and see how public participation affected the implementation of court orders.

2.   Case studies of judicial activism

Leather and leather goods is the eighth major export of India, amounting to approximately $19.10 billion in 2001–2 (Ministry of Finance, 2002). Small-scale units account for 60–65 per cent of the production and 25 million workers in total are engaged in this industry.[7]

The beginning of large-scale pollution discharge in this industry can be traced back to the change in the government's policy for the leather industry announced in the 1970s. Until then, raw skin and hide, as well as semi-processed goods, had been major export items. The new leather policy basically banned exporting these items and promoted leather goods in order to generate more foreign earnings from exports and create employment. As a result of the policy change, vegetable tanning, which requires more time, space and is more expensive, has been wiped out, and instead chrome tanning, which uses many chemical substances and materials, has became the major method of production. As has already been explained, most of the tanneries are small-scale and so have limited ability to install effluent treatment plants due to lack of financial and technical resources. It has been said that as many as 120 kinds of chemical substances have been discharged into local water sources, including rivers and groundwater. This has led to a decline of the agricultural crop and a deterioration of health among the local residents.

The government of India regards the leather industry as one of the 17 hazardous industries that are selectively monitored regularly. The National Environmental Action Plan describes the development of clearer production methods in tanneries as one of the priority programmes. The leather and leather goods industry is, therefore, regarded as being one of the top priorities of pollution control; however, at the same time, it is difficult to monitor all the units due to the large number of small-scale units, and it is also difficult to order non-complying units to close or terminate their operations due to their large contribution to employment and foreign earnings from exports.

Tamil Nadu

The state of Tamil Nadu is the biggest leather and leather goods production centre in India. The state's share of leather and leather goods exports is about 40 per cent and its share of production capacity is approximately 70 per cent.[8] Many industrial associations, such as All India Skin and Hide Tanners and Manufacturers Association and the Council for Leather Export, and research and development centres, such as the Central Leather Research Institute, are located in Chennai, the capital city of Tamil Nadu. Therefore, units in this state have a relative advantage in getting information, advice and help from associations and technical institutions.

The production of leather and leather goods in the state started in the nineteenth century; however, production increased after the declaration of the new leather industrial policy in the 1970s. Currently, the Vellore district of the state alone produces about 37 per cent of the total production in the country, 68 per cent of which is produced by micro-enterprises (less than 1,000 kg per day) (Kennedy, 1999). In that district, 25 tons of waste and 3,000 litres of water are discharged by the leather industry every day (*Frontline*, 8 October 1999). Since the 1980s, the areas along the Palar and Kauvery rivers have suffered from contamination of water, decline in main agricultural production and frequent occurrence of skin and liver diseases among local residents (Janakarajan, 2002).

In 1981, the state government ordered units to construct effluent treatment plants, with orders for small and medium units to jointly construct common effluent treatment plants (CETP). However, nothing developed until the central government announced a subsidy for the construction of CEPTs in 1989. Consequently, Tamil Nadu Leather Development Corporation constructed a CETP in 1991 (Sahasranaman, 2003).

In the meantime, the local residents have protested against pollution in various ways since the 1980s. In 1984, the state government formed a lawyers' committee to investigate pollution from the leather industry; however, the state government did not implement the recommendation of the committee or any other demand of the local residents. One of the committee members filed a petition seeking a moratorium on the issuance of licences by the state government to tannery units and asking that the state government take measures against pollution. However, the petition was put onto a pending list, so the lawyer formed a local NGO and filed a petition at the Supreme Court through PIL.

In 1992, the Supreme Court issued an interim order instructing the state government to provide victims with drinking water and medical relief. In 1995, the Supreme Court issued an interim order directing 57 units to close, while other units were directed to install individual effluent treatment plants or connect to CETPs within three months. In 1996, the Supreme Court issued another interim order commanding the shutdown of more than 200 units that had not installed any pollution control measures (*Frontline*, 11 August 1995; 31 May 1996). After these interim orders, the final judgement was issued in August 1996 (Vellore Citizens' Welfare Forum versus Union of India AIR 1996 SC 2715). Basically, the 'prevention of pollution' principle and 'polluter pays' principle were adopted. All units in five districts in the states were forced to pay fines of 10,000 rupees by the end of October and to install individual treatment plants or connect to CETPs by the end of November. The Ecology Authority was set up as an agency of the central government. After this judgement, the case was transferred to Madras High Court to follow up.

The Ecology Authority is mainly in charge of four areas as follows: (1) assessment of loss in the affected areas; (2) assessment of monetary loss in order to rehabilitate affected areas; (3) assessment of compensation for the victims; (4) monitoring pollution-discharging units; and (5) assessment of units located within 1 kilometre of rivers and relocation of them if necessary. The authority initiated these activities in 1998 and announced the names of victims and the amount of compensation for each of them in March 2001, while the High Court announced that units can pay by installment. The district collector is in charge of collecting compensation money from units and distributing it to the affected people (*Business Line*, 15 April 2002). Compensation will be distributed after 25 per cent of total amount (348 million rupees in total) is collected by units; however, only 13 per cent of the amount had been collected as of 2002 (*Frontline*, 16 August 2002).

Among the final orders of the Supreme Court, there was dispute over pollution control between the industry and local residents. Specifically, the parameter of 'total dissolved solid' was set at less than 2,100 mg per litre by the SPCB; however, up to 7,500 mg per litre was, in fact, approved by the state authority (*Hindu*, 3 April 2003). The industry considers the standards applied to it to be too strict compared with other states and other countries, and under the increasingly severe competition, enterprises, and especially small-scale units, insist that they cannot afford to pay more for pollution control. On the other hand, the local residents see some units illegally discharging effluent into rivers instead of connecting to CETPs, and residents insist that pollution still continues.

Kanpur and Kolkata

Kanpur, in Uttar Pradesh, is India's production centre for saddlery, harnesses, industrial boots and shoes, etc., while Kolkata, in West Bengal, mainly produces leather gloves and other leather goods. There are 392 factories in Kanpur and 538 in Kolkata (Sahasranaman, 2003), which make these areas the third largest and second largest, respectively, in terms of the number of factories.

In these two cities, the national project called Ganga Action Plan has much to do with pollution control since both cities are located on the Ganga River. M. C. Mehta, a Supreme Court advocate, filed a case through PIL in 1985 where he stated that the leather industry should close or should install pollution equipment. In 1988, the Supreme Court ordered that leather units must take measures or close (M. C. Mehta versus Union of India, Ganga Pollution [Kanpur Tanneries] Case AIR 1988 SC 1037). After this order, approximately 250 units eventually closed and, at the same time, 300-odd units linked to CETPs, supported by Dutch aid under the Ganga Action Plan. Furthermore, a new CETP was constructed when approximately 20 companies commenced new units in Unnao, a city adjacent to Kanpur. The quality of water, however, has not improved dramatically. The BOD figure for water discharged by units with individual effluent treatment plants is three times higher than the standard

(Alam, 2002). A local environmental NGO also reported that one third of the effluent runs directly into an irrigation canal, while two thirds runs into the Ganga River without any treatment, due to one of the channels being in need of repair (Toxics Link, 2001). As a result, the production of cash crops, particularly roses, declined and local residents suffered from skin and stomach diseases in the surrounding 20 villages.[9]

M. C. Mehta also filed a petition through PIL in Kolkata, which is located at the mouth of the Ganga River. In 1993, the Government of West Bengal reported that the state government would construct a leather complex, including an effluent treatment plant, and shift all the leather units in and around the city to the complex. After several interim orders, in 1997 the Supreme Court issued an order as follows (M. C. Mehta versus Union of India [Calcutta Tanneries] 1997 [2] SCC 411): (1) units must relocate to the complex specified by the government, otherwise they must shut down; (2) units must pay 25 per cent of land fees as deposits for operation in the complex to the state government, otherwise they must shut down; (3) each unit must pay 10,000 rupees as penalty; (4) the state government and High Court must use the penalty money for restoration of the affected environment and conservation; and (5) units must guarantee employment even after the relocation of units, and pay compensation for the relocation of employees. It has also been announced that the case was transferred to the Kolkata High Court to follow up. The opening of the leather complex is delayed. In the meantime, the authorities have disconnected water and electricity connections to units as per the court order, and the units operate by using handpumps and drawing electricity illegally (*Times of India*, 22 November 2002).

3. The role of community under judicial activism

Advantages and disadvantages of judicial activism

The three cases show that PILs did play a significant role in installing effluent treatment plants in tanneries. Court orders often specify the name of the companies that have to shut down or install effluent treatment plants, etc., and their time limit. However, it is not always the case that the installation of effluent treatment plants leads to less pollution emission.

As I have already shown in the three case studies, the advantages of judicial activism are its creation of awareness about pollution, sometimes through the media, and its effect of making the government administration and units enforce and abide by environmental laws and regulations. On the other hand, there are six main limitations to judicial activism. First, PIL takes a shorter time than other formal legal procedures; however, it still takes time. It took five years in the Tamil Nadu case and three years in the Kanpur case from the time the PIL was filed. In the meantime, non-complying units could continue operating. Second, any concerned person can resort to PIL. Even those who have little to do with the affected area can file a case under PIL. Consequently, the judgements might be in favour of complaints, but not necessarily in favour of those who are affected by the pollution. Third, technical aspects are not discussed very much. For example, there is no proof that the adoption of CEPTs is the best solution technically. Fourth, judgement should be rational and not only based on environmental protection, but also based on socio-economic conditions in the affected area. Employment is one of the key issues when units are forced to relocate to another area. In the above Kolkata case, labour and labour unions demand proper rehabilitation (*Times of India*, 19 January 2003), and there is no facility for workers and their families to stay in or around the new complex (*Times of India*, 20 January 2003). There is a possibility that many of the workers will lose their employment or will have to work without any compensation for the relocation. Fifth, there are no clear links between each judgement and national policy. Basically, each judgement provides remedy within the framework of command and control regulation. Economic incentives are often not considered. Finally, the administration is responsible for the implementation of orders. Judicial activism sometimes grants more power or budget to the administration; however, strict judgements do not lead to strict enforcement at units because SPCBs are not empowered and their weak monitoring and inspection capacities have not been strengthened, as I have repeatedly noted.

In each court case, the construction of CETPs has been promoted; however, the quality of water has not improved significantly. The possibility that units do not always operate effluent treatment is quite high.[10] That means that court orders can force the installation of treatment plants somehow, but they cannot force units to operate them. I will discuss how units could be made to operate treatment plants, particularly by looking at the community's role as a watchdog.

The community as watchdog

All three cases show that implementation of strict judgements is slow. Relatively speaking, Tamil Nadu is said to be much better in implementing court orders and pollution control in general than the other two states. For example, 'Almost all the facilities are covered by chrome recovery units, and this is not the case with Kanpur and Kolkata' (*Down to Earth*, 15 December 2002). The percentage of connections to CETPs among small and medium enterprises is as high as 80 per cent in Tamil Nadu (Kennedy, 1999). The United Nations Industrial Development Organization (UNIDO) expanded pollution control in the tannery industry in Southeast Asia, based on the experiences of Tamil Nadu. Supposing that Tamil Nadu is a 'successful' case, or at least a little better than the other cases, what are the differences between Tamil Nadu and other states?

Alam (2002) pointed out in his Kanpur tannery case study that the pressure from SPCBs is the most important motivation for units to implement pollution control currently, as well as to introduce emission treatment in the future. Considering the size and geographical spread of the industry, it is difficult to find out if the Tamil Nadu SPCB is far better than the other two SPCBs in Table 7.7.

Table 7.7   Performance of state pollution control boards

	Estimated no. of polluting units	No. of technical staff in position	No. of technical staff/no. of polluting units (estimated per 100 units)	Total revenue during 8th five-year plan (Rs. Lakh)	Training expenditure as a per cent of total expenditure	Total expen-diture	Surpluses as a per cent of total revenue	No. of regional and sub-regional offices	No. of regional, sub-regional, and mobile labs
Tamil Nadu	8151	295	3.62	5889.02	6.96	3894.28	44.89	25	14
Uttar Pradesh	6441	199	3.09	4562.74	2.37	2548.38	44.15	16	13
West Bengal	3414	85	2.49	1606.39	1.62	762.00	52.56	4	2
All Boards	–	–	3.7	–	–	27194.60	–	–	–

Source: Planning Commission, 2000.

In particular, the number of technical staff involved in monitoring and inspection is quite similar, at approximately three technical staff per 100 polluting units in the three states. Kennedy (1999) analysed the process of installing CETPs in Palar Valley in Tamil Nadu from the standpoint of collective action of industrial clusters and reported that some CETPs are operating efficiently and effectively. It is evident that industrial units can treat effluent even with weak enforcement of regulations.

So what is the missing link between the weak SPCBs and units' connection to CETPs? Kennedy (1999) found that social ties and cooperation in industrial clusters led to the operation of CETPs in order to avoid a bad personal reputation in the larger community. Since industrial associations and research institutes are mainly located in Tamil Nadu, tanneries in Tamil Nadu must have a comparative advantage in access to information, technology and other resources. Apart from these advantages, there is also a distinct difference between the case of Tamil Nadu and the other cases. In the Tamil Nadu case, the complainant is a local NGO. In addition to this NGO, there are several other NGOs that have dealt with this particular pollution problem by protesting, holding workshops, holding public discussions, appealing to the media and so on since the 1980s. As a result of these protests, the community finally resorted to the judiciary in order to seek help and solve the problem. Furthermore, there was a series of stakeholder meetings, which were chaired by a local research institute, in order to promote dialogue among local residents, NGOs, farmers' representatives, industry and government. Among these participants, there were differences of opinion, with some demanding complete closures of tanneries in the district and others simply wanting the river cleaned by operating the treatment plants properly, among other opinions. Anyway, in the other two cases, it was rare that local people voiced their opinions in front of industries; in Tamil Nadu, they know which units actually discharge effluent at night, discharge more than they are permitted to discharge, or illegally discharge into local water bodies.

The two cases in Kolkata and Kanpur were filed by a prominent environmental advocate. The relationship between the complainant and those who were affected was different from that of Tamil Nadu. In the case of Kanpur, those who were affected by tannery effluent could not get any compensation because compensation was not raised as an issue in the court case. In the case of Kolkata, since those who work in the tannery hail mainly from poor families of a specific low-ranking caste, who migrated from other areas, the local residents' attention to the problem is very limited. It is obvious that local residents in Vellore district of Tamil Nadu are more concerned about tannery pollution than those in the other two cities, although the Vellore residents' educational level and income are not really higher than those in the other two cities.[11]

There is an environmental NGO in Kanpur that made efforts to fight the pollution of the Ganga River and filed a case at High Court (Rakesh Kumar Jaishwal versus State of Uttar Pradesh and Others). This court case received much local attention due to media reports; however, once the judge who initiated the case retired, the case became low profile. A court-led case can attract people's attention for a short while, but the attention does not last long. More importantly, in the long run, such cases do not promote participation by a wide range of people in pollution monitoring and inspection. In fact, Dembowski (2001) notes that pressure groups in civil society will not be nurtured if the judiciary can solve the environmental problems effectively.

It is often reported that water pollution has not improved significantly despite the construction of CETPs. Needless to say, it is the state governments that monitor and inspect units in order to ensure that effluent treatment plants are operated properly and that effluent is not discharged illegally. Moreover, it is the local community that can supplement the government's role as a watchdog of both industry and government.

Conclusion

This chapter has examined judicial activism from the viewpoint of civil society's participation. It is indeed significant that the judiciary plays a role in pollution control; however, there are also shortcomings in judicial activism. Community participation under the judicial activism is often confined to using PIL. However, a community's informal monitoring and pressure does affect the implementation of the court judgements, not only in installing effluent treatment plants but also in operating them.

Since the 1990s, decentralization has been promoted by amendments of the constitution. With the approval of the state assembly, state governments can pass their responsibility for some matters, including environment-related matters, to local governments. This chapter demonstrates that the mere transfer of responsibility from the Supreme Court to a High Court or lower court, as well as from the central government to state government or district administration does not automatically solve the implementation problems. The community that stimulates the administration and functions as watchdog is critical, and it should be stressed that the success of decentralization depends upon how actively local communities function. Neither the improvement of court orders nor the empowerment of the lower part of the administrative structure will result in more accountable, responsible and transparent environmental problem-solving.

Notes

1. See Gupta and Goldar (2003).
2. In 1984, a poisonous chemical leak from the Union Carbide pesticide factory in Bhopal resulted in death of at least 16,000 people and maimed more than 50,000 people. This tragedy is regarded as the worst industrial disaster in history.
3. The seven priority areas are (1) conservation of biodiversity and the sustainable use of resources; (2) securing unpolluted water sources; (3) pollution control and

reduction of wastes, especially hazardous waste; (4) encouraging the adoption of clean technologies; (5) initiatives on urban environmental problems; (6) better scientific understanding of environmental problems; and (7) development of alternative energy sources.

4. The government announced the introduction of a voluntary agreement, 'Charter on Corporate Responsibility for Environmental Protection' in 2003. See the Ministry of Environment and Forests (http://envfor.nic.in/) for more details. Also see *Down to Earth*, 23 April 2003, which critically examined the charter.
5. The definition of small-scale industry undergoes changes over the years in terms of investment limit. As of 2001, a small-scale industry is any industrial undertaking in which investment in fixed assets of plant and machinery, whether held on ownership terms of lease, hire, or purchase, does not exceed 10 million rupees.
6. The Public Grievances Office in the Ministry of Environment and Forests received 390 complaints in 2002/3, of which approximately 30 per cent were industrial pollution-related issues, according to an interview with Ministry officials in December 2003.
7. Indian Leather Industry Website (http://www.indianleahterportal.com/leather-industry-overview/).
8. See http://www.tntdpc.com/tsmes/index.php
9. Interview with Rakesh Jaiswal, EcoFriends (environmental NGO in Kanpur) on 11 December 2003.
10. Lindstad (1999) describes how pesticide factories in Gujarat, India, simply pay bribes or operate treatment plants only at the time of inspection.
11. The literacy rate is 73 per cent in the Vellore district of Tamil Nadu, 78 per cent in the Kanpur Nagar district and 81 per cent in Kolkata, according to the 2001 Census.

References

Agarwal, Anil, Sunita Narain and Srabani Sen (eds) (1999) *State of India's Environment: The Citizens' Fifth Report*, New Delhi, Centre for Science and Environment.

Alam, Ghayur (2002) *Barriers and Opportunities for Promoting Trade in Environmentally Friendly Products – A Study of India's Leather Industry*, prepared for Consumer Unity Trust, Jaipur, and United Nations Environmental Programme.

Central Pollution Control Board (2001) *Pollution Control in Small-Scale Industries: Status and Needs*, Ministry of Environment and Forests, December.

Central Pollution Control Board (2003) *Assessment of Industrial Pollution*, Programme Objective Series: PROBES/92/2002–03, Ministry of Environment and Forests, March.

Confederation of Indian Industry (2002) *Compendium of Donor Assisted Projects in the Environment Sector in India*, compiled by the Confederation of Indian Industry for the World Bank, May.

Dembowski, Hans (2001) *Taking the State to Court: Public Interest Litigation and the Public Sphere in Metropolitan India*, New Delhi, Oxford University Press.

Divan, Shyam and Armin Rosencranz (2001) *Environmental Law and Policy in India: Cases, Materials and Statues*, 2nd edn, New Delhi, Oxford University Press.

Gupta, Shreekant and Bishwanath Goldar (2003) 'Do Stock Markets Penalise Environment-Unfriendly Behaviour? Evidence from India', Working Paper No. 116, Centre for Development Economics, Department of Economics, Delhi School of Economics.

High Powered Committee on Management of Hazardous Wastes (2001) 'Report of the High Powered Committee on Management of Hazardous Wastes Volume 1', Chairman: M. G. K. Menon, Writ Petition No. 657/95, Research Foundation for Science, Technology and Natural Resource Policy versus Union of India and Others.

Janakarajan, S. (2002) 'Soiled Agriculture and Spoiled Environment: Socio-Economic Impact of Groundwater Pollution in Tamilnadu', Working Paper No. 175, Madras Institute of Development Studies.

Kennedy, Lorraine (1999) 'Cooperation for Survival: Tannery Pollution and Joint Action in the Palar Valley (India)', *World Development*, 27(9), 1673–91.

Kuik, O. J., M. V. Nadkarni, F. H. Oosterhuis, G. S. Sastry and A. E. Akkerman (1997) 'Pollution Control in the South and North: A Comparative Assessment of Environmental Policy Approaches in India and the Netherlands', *Indo-Dutch Studies on Development Alternatives*, 21, New Delhi, Sage Publications.

Lindstad, Petter (1999) 'International Production of Pesticides: Case Study of Gujarat, India', in Stig Toft Madsen (ed.), *State, Society and the Environment in South Asia*, Nordic Institute of Asian Studies, Man and Nature in Asia Series No. 3, Surrey, Curzon.

Ministry of Environment and Forests (1992) *Policy Statement for Abatement of Pollution*, Government of India.

Ministry of Environment and Forests (1993) *National Environmental Action Plan: Environment Action Programme*, December, Government of India.

Ministry of Environment and Forests (2003) *Annual Report 2002/03*, Government of India.

Murthy, M. N. and U. R. Prasad (1999) 'Emissions Reduction and Influence of Local Communities in India', M. N. Murthy, A. J. James and Smita Misra (eds), *Economics of Water Pollution: The Indian Experience*, New Delhi, Oxford University Press.

Nomura, Y. and T. Endo (1994) 'India's Environmental Laws and Administrative System', in Y. Nomura and N. Sakumoto (eds), *Environmental Laws in Developing Countries: South East Asia and South Asia*, Tokyo, Institute of Developing Economies.

Pargal, Sheoli, Mani Muthukumara and Mainul Huq (1997) 'Inspections and Emissions in India: Puzzling Survey Evidence about Industrial Pollution', Policy Research Working Paper 1810, World Bank.

Planning Commission (2000) *Evaluation Study on Functioning of State Pollution Control Boards*, Government of India.

Planning Commission (2001) *Indian Planning Experience: A Statistical Profile*, Government of India.

Prasad, P. M. (2004) 'Environmental Protection: The Role of Liability System in India', *Economic and Political Weekly*, 17 January, 257–69.

Sahasranaman, A. (2003) *Environment Management: A Study of the Indian Leather Industry*, Council for Leather Exports.

Sankar, U. (2001) *Economic Analysis of Environmental Problems in Tanneries and Textile Bleaching and Dyeing Units and Suggestions for Policy Action*, New Delhi, Allied Publishers.

Sato, Hajime (2001) 'Public Interest Litigation in India: An Analysis of Remedies and Procedures in the "New" Litigation (I)' (in Japanese), *Ajiakeizai*, 42(6), 2–25.

Sawhney, Aparna (2003) 'Managing Pollution: PIL as Indirect Market-Based Tool', *Economic and Political Weekly*, 4 January.

Toxics Link (2001) *Pipe Dreams Common Effluent Treatment Plants: Technology, Management and Experience*, Toxics Link.

World Bank (1995) *India's Environment: Taking Stock of Plans, Programs and Priorities, An Assessment of the Environment Action Program*, India, Nepal, Bhutan Country Department, South Asia Regional Office, World Bank.

World Bank (2000) *Greening Industry: New Roles for Communities, Markets and Governments*, New York, Oxford, World Bank.

World Wild Fund for Nature India (WWF India) (1999) *Environmental NGOs in India: A Directory*, 7th edn, Indira Gandhi Conservation Monitoring Centre, WWF India for the Environmental Information System (ENVIS) of the Ministry of Environment and Forests, Government of India.

8
Environmental Policy under Multi-stakeholder Governance in China: Focusing on Implementation of Industrial Pollution Control

Kenji Otsuka

Introduction[1]

China has developed its environmental policy by mainly focusing on industrial pollution control for 30 years. In the beginning, environmental policy in China was initiated by the central government under the leadership of top leaders in the Communist Party, and administrative and law systems have been strengthened gradually. Since the 1990s, the State Council, the highest administrative body in China, has called all industries to comply with emission standards and has also issued implementation guidelines to order local governments to shut down small-scale industries that cause heavy environmental pollution. On the other hand, the government has cooperated with the newly reformed People's Congress and the mass media, even though all media in China are still under the control of the government and the Communist Party, to supervise and inspect the implementation process of strict command and control regulations. At the same time, in recent years NGOs and residents are also expected to play important roles in implementing the regulations.

As stated in Principle 10 of the Rio Declaration, which was adopted at the 1992 UN Conference on Environment and Development,[2] 'Environmental issues are best handled with the participation of all concerned citizens, at the relevant level.' Broader stakeholders' and citizens' participation is widely considered to be inevitable and effective in environmental governance, not only among developed countries but also among developing countries. Although China is no exception to the trend towards building multi-stakeholder governance in implementing environmental policy,[3] its effectiveness is inclined to be questioned due to the fact that it is a socialist state under control of the Communist Party.

When discussing the effectiveness of multi-stakeholder governance for environmental policy in China, one should be aware not only of the development process in environmental policy itself, but also of its interplay with socio-economic institutions under economic, social and political reforms over the course of 20 years. Analysis of the environmental policy process from this point of view is essential. This chapter will review multi-stakeholder governance in environmental policy implementation, focusing on industrial pollution control in China in an analysis of the policy process through official documents, newspaper articles, interviews with stakeholders and other first-hand materials that the author has obtained through a series of cooperative research projects with Chinese research institutes and scholars.[4] Through this analysis, the author will test the effectiveness and the problems in the implementation of environmental policy under the current multi-stakeholder governance system in China.

The content of this chapter is as follows. In the Sections 1 and 2, the process of environmental policy development and social change in China for the past 30 years will be reviewed. In Section 3, new trends in information disclosure and public participation in implementing an environmental policy to strengthen the multi-stakeholder governance system in China will be described. In Section 4, characteristics, effectiveness and problems in multi-stakeholder governance in implementing industrial pollution control in contemporary China will be discussed. In the final section, Section 5, the author concludes with findings and prospective tasks.

1. Building the government initiative system during the 1970s and 1980s

Launching environmental policy

There emerged serious environmental pollution problems and even conflicts between polluting factories and damaged farmers because of unregulated economic development in China in the late 1970s, and central and local governments had to respond to these problems (ZHX, 1994, pp. 3–6). Although the view that environmental problems could not occur in socialist states was dominant in the era of the Cultural Revolution, official delegates were sent to the UN Conference on the Human Environment at Stockholm in 1972 at the initiative of Premier Zhou Enlei, who was anxious about the seriousness of environmental deterioration in China. This event pushed forward environmental policy development at the First National Conference of Environmental Protection held in China in 1973 (Qu, 1984, p. 109, 1997, Intro. p. 3).

An important juncture was reached in the field of environmental policy, as in many other policy fields, when the political line of the Chinese Communist Party (CCP) was established as 'Reform and Open Door' after the Third Plenum of the Eleventh Central Committee of CCP held in December 1978. It was also in December 1978 that CCP recognized the seriousness of environmental destruction and the needs of environmental policy in its official document, Central Committee Document No. 79, for the first time. Since then, roles of public opinion and mass media have been considered positive factors when they refer to the experiences of environmental policy in developed countries like Japan.[5] At the Second National Conference of Environmental Protection held from December 1983 to January 1984, it was declared that environmental policy, as well as family planning (the so-called 'one child policy'), were basic state policies and that it was essential to harmonize economic development with environmental protection (ZHX, 1994, pp. 20–7).

Regulation and administrative organization

At the beginning of building a governance system for environmental policy implementation in China, regulations, administrative organization, and a propaganda paper were set up as its key components.

Environmental regulations in China consist of a constitutional clause concerning environmental protection, legislation passed through the National People's Congress (NPC), administrative regulations enacted by the State Council, the National Environmental Protection Agency (NEPA)[6] and other divisions, and many local regulations (Kataoka, 1997). In China's Environmental Protection Law (tentative), enacted in 1979, it was stipulated that 'any citizen has the right to supervise, impeach, and accuse bodies and individuals that pollute and destroy the environment'. It was also stipulated that environmental protection agencies could charge polluters for damages. Under these stipulations, a citizen, upon suffering pollution harm, can impeach, accuse the polluter and demand compensation (Wang *et al.*, 2001, p. 5). It should also be noted that a decision by the State Council would be hierarchically assumed as lower than legislation by NPC theoretically, but that a series of the decisions have played an essential role in determining the direction of national environmental policies (see Table 8.1).

Administrative organization in central and local government is the key to enforcing environmental regulations.[7] Figure 8.1 shows the basic concept of environmental administrative organization in China. The environmental administrative system is composed of three areas, which are government, environmental protection administration and environmental protection offices in other administrative organs. First, the governmental system in China is constructed of five hierarchical levels, which are state, province, district and city, county and township and village. Under the principle of centralization, each local government should follow every policy of the State Council (Kobayashi, 1999, p. 127; Xie, 1998, p. 115). According to the Environmental Protection Law, revised in 1989, each local government should be responsible for the quality of its own environment.

The second area in the environmental administration system is the environmental protection administration in central and local government. The National Environmental Protection Agency was set up in 1984, and its functions were strengthened as an independent organization under the direct control of the State Council in 1988.[8] The Environmental Protection Bureau (EPB) was also set up at local levels. According to the Environmental Protection Law, revised in 1989, NEPA should implement national environmental

Table 8.1   Major regulations for environmental pollution control in China

Month/year of promulgation	Year of revision	Name of regulation
9/1979	9/1989	Environmental Protection Law
2/1981	–	State Council Decision on Enhancement of Environmental Protection Works
2/1982	7/1998; 1/2003	Measure for Collection of Pollutant Emission Charges
8/1982	12/1999	Marine Environmental Protection Law
5/1984	5/1996	Water Pollution Control Law
5/1984	–	State Council Decision on Environmental Works
3/1986	11/1998	Measure for Environmental Management of Construction Projects
9/1987	8/1995; 4/2000	Air Pollution Control Law
9/1989	–	Noise Pollution Control Law
12/1990	–	State Council Decision on Further Enhancement of Environmental Protection
12/1990	4/1997	Measure for Environmental Claim Management
7/1991	–	Tentative Measure for Environmental Administrative Works
7/1992		Measure for Punishment for Environmental Protection
8/1992		Tentative Measure of Certification for Environmental Administrative Law Enforcement
8/1992	–	Ten Major Policies for Environment and Development (approved by CCP and State Council)
8/1995	–	Tentative Regulation for Water Pollution Control in the Huai River
10/1995	–	Solid Waste Pollution Control Law
8/1996	–	State Council Decision on Some Problems Concerning Environmental Protection
10/1996	–	Noise Pollution Control Law
10/2002		Environmental Impact Assessment Law
6/2003		Radioactive Pollution Control Law

protection policy, while EPB, as its counterpart in local government, should coordinate and manage environmental policy under its jurisdiction. Through this chain of command, the upper EPB directs the lower EPB (Cheng and Piao 1994, p. 41; Kataoka, 1997, p. 274).

The third area consists of the Environmental Protection Offices (EPO) under other administrative organs. It should be noted that the environmental

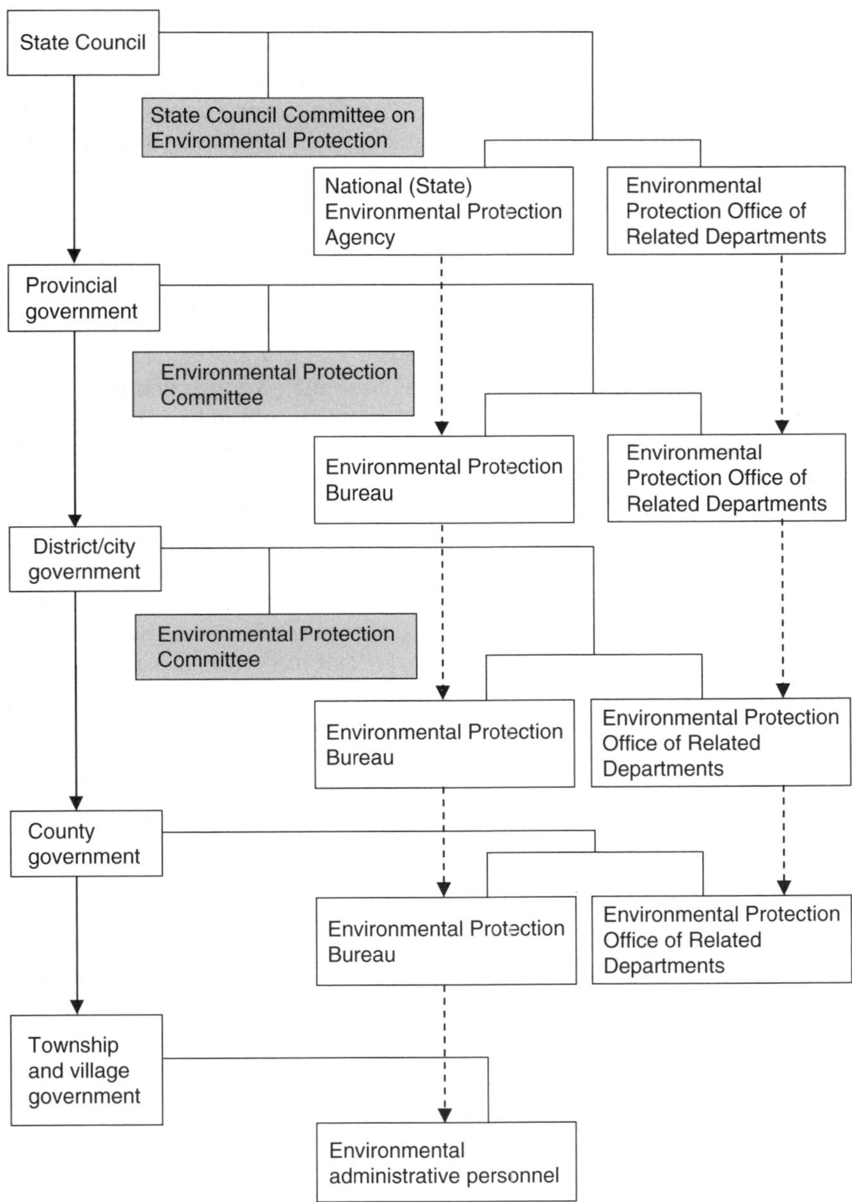

Figure 8.1 Environmental administrative organization in China
Note: The State Council Committee on Environmental Protection was abolished and the National Environmental Protection Agency was reformed as the State Environmental Protection Agency in 1998.

protection offices are not 'directed', but 'supported' by the Environmental Protection Bureau (Chen and Piao, 1994, p. 42).

To coordinate the second and the third areas of administration, the State Council Committee on Environmental Protection was established in 1984 and played important roles in decision-making and implementation of environmental policy until its abolishment in 1998. The top leader of the committee was a deputy prime minister or state councillor, and its members, up to nearly 50 at its peak, were mostly deputy directors of each appointed division. The committee was held almost every four months and discussed important national environmental policies. It also reported that 23 provincial governments and 11 major cities had set up a committee on environmental protection like State Council's (ZHNBW, 1990, p. 79).

It was the State Council Committee on Environmental Protection that played a major role in building the government initiative system for implementing environmental policy in China. In the initial stage, from 1984–8, the committee held debriefing sessions from some deputy governors of provinces, cities and counties to review how well each local government was working in implementing national environmental policy. In the second stage, from 1988–93, the committee stepped deeply into the local implementation through on-site visits by its members, on-site meetings by central and local stakeholders and efforts to work out the pollution control programme. In Benxi City, Liaoning Province, in northeastern China, where large-scale iron, steel, cement, coal and chemical industries producing heavy air and water pollution were located, the committee declared its 'Decision on Environmental Pollution Control in Benxi City' in 1988, and approved the city's six-year programme on industrial pollution control as one of the national pilot projects. This was the first time the committee directed environmental policy implementation at the local level (GHWB, 1988; GHWM, 1995).

Propaganda and claim management

The third key component of the governance system for environmental policy implementation in China at its launch was the propaganda and claim management system. The propaganda paper called *China Environment News (Zhongguo Huanjing Bao)* has been published to propagate national environmental policy by the State Council Committee on Environmental Protection since 1984.[9] Although this newspaper has also been expected to arouse and absorb pubic opinion by reporting serious pollution accidents and typical cases of environmental deterioration, politically sensitive problems such as serious pollution disease have not been publicized in the paper but have been reported secretly to some limited leaders by the inner circular called 'Environmental Situations' (*'Huanjing Qingkuang'*) (ZHNBW, 1990, pp. 189–90). Also, people can complain about their concerns and issues relating environmental problems to administrative divisions as well as to this paper and other papers (*xinfang*). At the environmental administrative level, Measures for Environmental Claim Management were stipulated in 1990 and revised in 1997 (ZHNBW, 1998, pp. 45–7).

Some government leaders responsible for environmental policy, like Qu Geping, founding director of NEPA, and Li Peng, founding chairman of the State Council Committee on Environmental Protection, were aware of the important role of public opinion in enforcing environmental regulations (Qu, 1984, pp. 295–8; GHWB, 1988, pp. 38–40), and some cases where propaganda worked well for enforcing regulations were also reported in official meetings. However, China seemed to be reluctant to let mass media report a lot of serious problems that were under the strict control of the propaganda branch of Communist Party and government. Thus, it has been found that some leaders who were responsible for environmental policy were aware of the role of mass media and public opinion in strengthening policy implementation, but that role of the mass media was not enhanced until economic, social and political changes took place in the 1990s.

2. Transformation to multi-stakeholder cooperative system in the 1990s

Institutional development for environmental law enforcement

Once environmental laws were established, the Chinese government had to face the problem of how to enforce them effectively. The State Council issued its decision on Further Enhancement of Environmental Protection Works in December 1990 to demand that each local government, administrative division, enterprise and institution rigidly enforce environmental laws and related regulations and that each environmental protection and legal affairs division conduct routine inspection for law enforcement in order to improve situations such as 'no reliance on laws, not conforming to rules, not rigidly enforcing laws, not condemning illegalities, and using not laws but power' (ZHNBW, 1991, pp. 19–21).

After the first promulgation of the Administrative Procedure Act in China in April 1989, NEPA issued the Measures for Punishment for Environmental Protection in July 1992 (ZHNBW, 1993, pp. 8–10). NEPA also issued the Tentative Measures for Environmental Administrative Work in July 1991 and Tentative Measures for Certification for Environmental Administrative Law Enforcement in August 1992 to create permanent positions for environmental administrative personnel who are responsible for supervision and inspection of environmental law enforcement, supervision and inspection of marine environment and natural ecosystem preservation, investigation and administration of pollution incidents and disputes, and collection of pollution discharge fees (see also Table 8.1).

Figure 8.2 shows fluctuations in the number of local environmental administrative personnel. It tells us that personnel in Environmental Administrative Offices (EAO) have increased from about 7,000 in 1992 to over 30,000 in 2000.

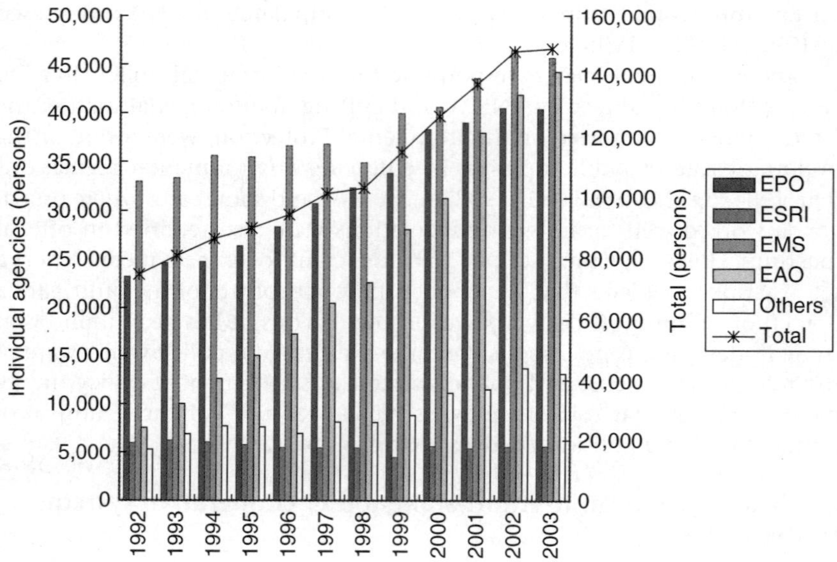

Figure 8.2   Numbers of local environmental administrative personnel
Notes: 1. These figures include provincial, district/city and county administrative personnel.
2. EPO: Environmental Protection Office; ESRI: Environmental Science Research Institute; EMS: Environmental Monitoring Station; EAO: Environmental Administrative Office.
Source: Compiled by the author based on *China Environmental Yearbook* (*ZHNBW, 1993–2002*).

Prioritizing sustainable development strategy

CCP held the Fourteenth National Congress in October 1992, where Secretary General Jiang Zemin, in his political report, addressed 'enhancement of environmental protection' as one of main tasks of open-door policy, reform and modernization of China during the 1990s. It should be noted that there emerged not only a sense of environmental crisis among CCP leaders but also a wish for international cooperation following the United Nation Conference on Environment and Development (UNCED) held in Rio in June 1992.[10]

After UNCED, in August 1992 the Central Committee of CCP and the State Council jointly issued Ten Major Policies for Environment and Development with the top priority on 'sustainable development strategy' for domestic policy implementation (ZHX, 1994, pp. 457–60), and the State Council published *China's Agenda 21*, supported by the United Nations Development Programme (UNDP) for the international community in 1994 (LGCA, 1994).

It was *China's Agenda 21* that was the first official document to stipulate roles in information disclosure and public participation, such as promotion of promulgation by mass media and public participation in the legislative, judicial and administrative processes.

Supervision and inspection system for environmental law enforcement

After the political report presented by Secretary General Jaing at the Fourteenth National Congress of CCP, the State Council Committee on Environmental Protection (CEP) held its Twenty-fourth Meeting in January 1993 to deliberate on and approve the Decision on Enhancement and Development of Inspection for Environmental Law Enforcement and Rigid Crackdown for Malpractices, and the State Council issued it in that March. What is important in this decision is the demand that the propaganda branch of Communist Party, the government and the mass media reveal names of organizations and persons who committed serious offenses against an environmental law (ZHNBW, 1994, pp. 9–10, 86).

It was also in 1993 that the Committee on Environmental Protection was established, and it was renamed as the Committee on Environment and Natural Resource Protection (CENRP) in the next year, in the National People's Congress. Qu Geping, founding director of NEPA, was appointed as head of the committee, and Xie Zhenhua, former deputy director of NEPA, was promoted to director of NEPA. This means that CENRP of NPC became independent from NEPA officially but kept a close relationship with NEPA through its personnel (ZHX, 1994, p. 444). Other members of the committee were five deputy heads and eleven commissioners. The main functions of the committee were not only to submit and deliberate on drafts of environmental and resource protection laws and related bills, but also to supervise enforcement of laws on environmental and resource protection in cooperation with the Executive Committee of NPC. The CENRP of NPC played an important role in the National Inspection for Environmental Law Enforcement and the Century Walk for Environmental Protection in China, both of which were launched in 1993.[11]

The National Inspection for Environmental Law Enforcement was implemented as one of major activities of the Third State Council Committee on Environmental Protection (State Council's CEP), which cooperated with the Executive Committee and CENRP of NPC for three years from 1993. The main targets of the inspection were enforcement of the Environmental Protection Law, Wildlife Conservation Law, Air Pollution Control Law, Water Pollution Control Law, Forest Law, Grassland Law, Water and Soil Conservation Law and so on (ZHNBW, 1994, pp. 87–8, 100–1, 106). Direction of local environmental policy implementation by the State Council's CEP, which had been a major role of the First and Second Committee, was institutionalized through the national inspection.

The Century Walk for Environmental Protection in China was a kind of environmental protection campaign to arouse public opinion through national and local newspapers, radio and TV and to urge that problems be solved. This campaign was coordinated by its Organizing Committee, which consisted of

CENRP of NPC as head, the Propaganda Department of CCP and the Department of Radio, Cinema and TV in the State Council as deputy heads and the central mass media, NEPA and the Department of Forestry as members. Major mass media participating in the Organizing Committee are the *People's Daily*, *Xinhuashe*, *Guangming Daily*, *Economic Daily*, *Science and Technology Daily*, *Legal Daily*, China Environmental News Co., China Central TV (CCTV) and China National Radio (CNR) (Qu, 1997, Intro., pp. 20–1; ZHNBW, 1994, pp. 10–12).[12]

At the Fourth National Conference of Environmental Protection, held in 1996, the state of three years' activities of the national inspection and campaign was reported by NEPA. The state of the national inspection was reported as follows. The national inspection covered 26 provinces, over 240 districts and counties, 566 enterprises and institutions; over 320 persons from NPC, State Council and local people's congresses participated in this inspection; province-level governments investigated and dealt with over 6,000 violations of laws related to environmental protection and over 100,000 violations of laws related to natural resource protection; inspection teams received over 2,600 letters, visits and telephone calls from residents; over 40 journalists in total accompanied the inspection teams to publicize over 500 news articles (NEPA, 1996, pp. 318–26). The state of the campaign was as follows. Over 200 journalists from 22 mass media in Beijing publicized nearly 1,000 news articles; over 1,500 journalists from over 750 local mass media publicized over 15,000 news articles; among which reports on TV comprised 15 per cent, which adds up to over 1,600 TV reports (NEPA, 1996, pp. 335–6).

The most important achievement of the inspection and campaign would be to reveal serious environmental disruption and poor environmental management by local governments and firms nationwide. After the end of the three-year national inspection, the Executive Committee and CENRP of NPC have continued inspections on environmental law enforcement and publicized their reports yearly to promote new legislation and revision of environmental laws (Chen and Piao, 1997, p. 245; see also Table 8.1). Also, other types of inspection by governmental sectors and campaigns by mass media have been conducted up to now.

Furthermore, it should be noted that this series of activities has mobilized not only governmental organizations but also the People's Congresses, mass media and the public widely for environmental policy enforcement. The Decision of the State Council on Some Problems Concerning Environmental Protection was issued by the State Council after this conference, and it declared:

> A public participation mechanism should be set up, roles of social associations should be enhanced, public participation in environmental protection activities should be promoted, and illegal actions against environmental laws and policies should be brought to trial. The mass media, such as newspapers, radio and TV, should report timely typical advanced examples of

environmental degradation and also disclose the organizations and persons who cause serious pollution and ecological degradation, and this supervisory role of the mass media should be enhanced. (NEPA, 1996, p. 41)

That is to say, the State Council has officially admitted that mobilization of the mass media and promotion of public participation are inevitable measures for environmental policy implementation. Thus, China has stepped towards building a multi-stakeholder governance system in implementation of environmental policy.

Enhancement of industrial pollution control

At the start of the environmental protection campaign, CCTV revealed very serious water pollution problems in the Huai River in October 1993 as follow. All fish and shellfish were dead due to effluent from pulp plants and other industries; the mortality from cancer and the rate of birth defects among the residents who were drinking the river water were higher than those of others; and local and central government had received appeals from the residents along the river to fix the problems, but had not taken any effective action. The *People's Daily* and *China Youth Daily*, each of which are sponsored by CCP and the Communists' Central Youth Organization, also revealed a large-scale water pollution incident in the Huai River that occurred in July 1994,[13] in which nearly 1.5 million residents along the river could not obtain sufficient drinking water.

Through this series of events, the State Council and its Committee on Environmental Protection embarked on tougher regulation of industries placing a heavy water pollution load on the Huai River. The State Council issued the Tentative Regulation of Water Pollution Control in the Huai River in August 1995 and approved the Ninth Five-year Plan (1996–2000) of Water Pollution Control in the Huai River in June 1996. Under the regulation and the plan, small chemical pulp plants with less than 5,000 tons annual production were ordered to be shut down by 30 June 1996, and all manufacturing industries were required to comply with the effluent standard in this river basin by the end of 1997.[14]

Through the Decision of the State Council on Some Problems Concerning Environmental Protection issued by the State Council in August 1996, this Huai River pollution control model has been expanded nationwide. First, each local government and firm was ordered to ban operation of, close facilities of, or stop running 15 categories of small industries by 30 September 1996. These industries are chemical pulp, tanning, dye, cokes refining, sulphur refining, arsenic refining, mercury refining, zinc and lead refining, oil refining, gold mining and refining, pesticide, bleaching and dyeing, electroplating, asbestos and radioactive materials industries. Second, all polluting sources of industries were required to comply with national and local effluent standards by the end of 2000. Third, major polluted rivers and lakes, the so-called 'Three Rivers and

Three Lakes' (*san he san hu*), which are Huai River, Hai River, Liao River, Lake Tai, Lake Dianchi and Lake Zao, have adopted a policy measure of total COD emission control within each watershed. NEPA and the Ministry of Inspection inspected the enforcement of these regulations (NEPA, 1996, pp. 37–41).

3.   New trends for multi-stakeholder governance: promotion of information disclosure and public participation in implementing environmental policy

Promotion of information disclosure

As stipulated in Principle 10 of the Rio Declaration,[15] people's access to information and information disclosure by states is considered to be a key in promoting public participation in environmental policy. Besides environmental campaigns by the mass media as mentioned above, broad information disclosure to the public initiated by government is seen as one of the most important measures in environmental policy implementation in China in recent years.

Since 1997, major cities have launched an Air Pollution Index (API) report, which indicates the harmfulness of air pollution exposure to human heath by using five categories (or a total of seven categories when two subcategories are included), weekly or daily through the mass media and the Internet. In fact, the Beijing Municipal Government urged a series of strict air pollution control measures after weekly API reports from the mass media revealed serious air pollution in the city since 1998 (Otsuka, 2002b; Wang *et al.*, 2002, pp. 96–8). Thus, this practice of API reports is considered to be sensitizing not only the public but also the government.

Another form of information disclosure for environmental policy implementation is the Industrial Environmental Performance Disclosure Programme led by the World Bank, SEPA and some research institutes and local governments in China since 1998 (Wang *et al.*, 2002, pp. 122–200). The programme is the Chinese version of a programme called PROPER (Programme for Pollution Control, Evaluation and Rating) initiated in Indonesia by the World Bank (World Bank, 2000, pp. 57–79).[16] Table 8.2 shows the rating criteria of Zhenjiang City in Jiangsu Province, where the World Bank, the local government, the Institute of Environment and Nanjing University demonstrated the programme. As shown in this table, each firm is rated by five colours (categories) based on its environmental performance, such as emission standards achievement, causation of illegal acts or pollution accidents, ISO14000 and cleaner production. The rating of each firm is announced via local TV, radio and newspapers after the ratings are released to the press by city government. Through public disclosure of the ratings, this programme is expected to make local firms adopt cleaner technologies.

Based on the trial programme in Zhenjiang city, the government of Jiangsu Province has promoted this rating programme all over the province. It is conducted now not only in cities, but also in some villages. SEPA is also

Table 8.2   Rating criteria of PROPER in Zhenjiang City

State of achievement	Class	Behaviour of firm
Not achieved	Black	No endeavour for pollution control, causing serious environmental disruption
	Red	Some endeavour for pollution control, but has not achieved national pollution control standard
Warning	Yellow	Achieved national pollution control standard. However, has not achieved total emission control standard, or has been punished, or has caused a pollution incident
Achieved	Blue	Achieved local pollution control standard and basic needs of environmental management
	Green	Adopting cleaner production technology, acquired ISO14000, and performance has achieved international or advanced domestic standards

Source: Wang *et al.*, 2002, p. 160, Table 9.1.

now promoting this rating programme in several provinces and cities across the country.[17]

NGO participation in environmental law enforcement

Since CCP launched the Open-door and Reform Policy in the late 1970s, NGOs with a variety of organizational forms have been expanding their areas of activities. The groups considered to constitute the NGO sector in China are composed of the following four types: (1) social association-type NGOs; (2) service unit-type NGOs; (3) grassroots NGOs; and (4) international NGOs. With the exception of the social association-type and service unit-type groups, NGOs in China cannot register themselves under the present management system. Those that are not registered under the present system are required to be authorized by the government in one form or another. This is required in a political environment in which the party and the government find it necessary to strictly regulate anti-government and anti-social activities for the sake of the major principle of maintaining a single-party dictatorship. There are emerging grassroots NGOs that have been established by volunteers and that draw a line of demarcation between themselves and party-related or government-related organs, and they have been conducting environmental campaigns for several years; however, under this system and in this political environment, these NGOs are reluctant to advocate measures in direct contradiction to national policy, and they also face a disadvantage in mobilizing human and financial resources compared to other types of groups (Otsuka, 2002a).

Environmental protection is one of most active areas for NGOs in China. Leading environmental NGOs in China, especially those categorized as grass-roots NGOs, mentioned above, are Friends of Nature (founded in 1994) and the Global Village of Beijing (founded in 1996). These two NGOs are carrying out activities not only in the area of environmental education and non-governmental international exchange but also in introducing society at large to problems concerning environmental policy. Both were established by groups of intellectual volunteers and depend mainly on foreign funds. These organizations are still small in number, but they are the closest groups in China to the concept of NGOs as defined in the introduction to this book. They are antithetical to the social association-type NGOs, which are most common in China. Furthermore, informal activities and salons without permanent organizational form have begun to appear.[18]

In the context of the main theme of this chapter, we will focus on a unique environmental NGO struggling for better environmental policy implementation through its legal aid programme to pollution victims. It is called Centre for Legal Assistance to Pollution Victims (CLAPV), and it was registered with the Ministry of Law with permission from China University of Political Science and Law in October 1998.[19] CLAPV is run by professional volunteers, such as scholars and graduate students whose majors are environmental law, both in and outside the university, and it is supported by lawyers and journalists who are enthusiastic about environmental protection and is also sponsored by foreign foundations. The Director of CLAPV, Mr Wang Canfa, is a university professor of environmental law as well as an environmental lawyer (Wang *et al.*, 2002).

CLAPV launched a free telephone service for pollution victims in November 1999 and offers legal and financial aid to pollution victims nationwide. CLAPV has received visits from 333 persons in total, handled 4,282 phone calls and received 186 letters as of the end of 2002, and it has supported over 57 lawsuits as of 2004.[20]

Since 2001, CLAPV has focused on environmental disputes in western China where economic conditions are poor. According to the vice-director of CLPAV, Ms Xu Kezhu (2004), CLAPV has supported 12 cases, of which ten cases have been filed. These cases include air and water pollution damage not only to agricultural and aquacultural products, but also to human health due to emissions of pollutants from industries. They also include two class-action suits and one unfiled case in which over 10,000 persons claim to be pollution victims. However, CLAPV won only one case and agreed to settlements in two cases. The reasons for the unsuccessful cases are considered to be the intervention by local government favouring economic growth, a lack of understanding concerning environmental law enforcement among court justices, the lack of a comprehensive environmental disputes settlement system in the legal assistance process for pollution victims and so on.

Although CLAPV is confronted with many difficulties such as those above, it is expected to play a significant role in improvement, not only of the way in which pollution disputes are settled, but also in the way that environmental law is enforced in China. To augment the effectiveness of environmental law enforcement at the grassroots level, CLAPV has also organized a training course for local lawyers and justices once a year since 2001, in cooperation with SEPA, China Association for Lawyers and State Academy for Justices. The government and the legal community will support such training courses because they are fully aware that China needs to reform its legal system from the 'rule of men' to a 'rule of law' in order to live in the global market economy.[21] The training course is considered to be a boost for CLAPV in getting wider support from legal professionals at the grassroots level.

4. Discussion: effectiveness of industrial pollution control under the multi-stakeholder environmental governance in contemporary China

Problems of regulatory enforcement

It is clear that the regulations enhanced since the 1990s have urged the government and the industrial sector to control pollution sources to a certain degree. Both Figure 8.3 and Table 8.3 are based on statistics of *China Environmental Yearbooks*. Figure 8.3 shows the trends of investment in industrial

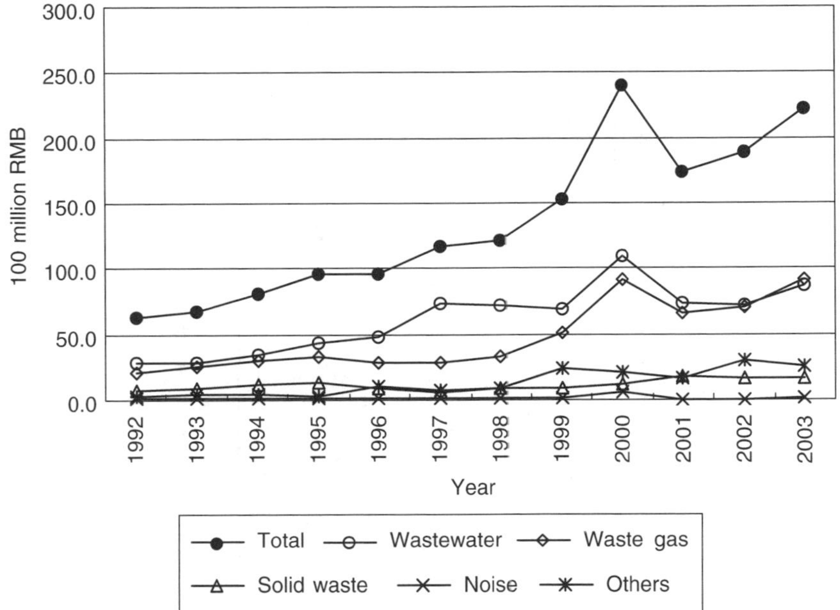

Figure 8.3   Investment in industrial pollution control
Source: Compiled by the author, based on *China Environmental Yearbook (ZHNBW, 1993–2000)*.

Table 8.3 State of wastewater and gas emissions from major industries

Year		1991	1992	1993	1994	1995	1996	1997	1998	1999	2000
Industrial wastewater	Total (10 million tons)	236	234	220	216	222	206	188	171	161	153
	COD (10,000 tons)	–	715	622	681	768	704	665	510	692	662
	Disposal ratio of waste water (%)	63.5	68.6	72.0	75.0	76.8	81.6	84.7	88.2	91.1	95.0
	Achievement ratio of wastewater discharge	50.1	52.9	54.9	55.8	55.4	59.1	61.8	67.0	72.1	82.1
Industrial waste gas	Powder dust (10,000 tons)	579	576	617	583	639	562	548	506	458	404
	Flue dust (10,000 tons)	845	870	880	807	838	758	685	680	557	517
	SO_2 (10,000 tons)	1165	1323	1292	1341	1405	1364	1363	1210	1078	1172

Note: – indicates no available data.
Sources: Compiled by the author, based on *China Environmental Yearbook (ZHNBW, 1992–2001)* and related articles in *China Environmental News*.

pollution control by major industries. From this figure, we see that total pollution control investment has increased rapidly since 1996, among which water pollution control investment is the leading sector. This statistical trend corresponds with the policy trend in which pollution control was mainly focused on industrial water pollution control in the late 1990s. Table 8.3 summarizes the state of wastewater and gas emission from major industries from 1991 to 2000. We see a trend of improvement in the achievement ratio of wastewater discharge, as well as a trend of decrease in total wastewater discharge and powder and flue dust emission over these ten years. However, there is no clear improvement in some figures, such as COD and SO_2 discharge.

Also, some problems are found in the process of regulatory enforcement. First, it should be pointed out that some industries could escape the regulations when they faced difficulties in achieving the regulatory demands, but others could not. For example, some large-scale and central state-owned enterprises admitted in January 2001 to being allowed to escape compliance with emission standards for a year by the State Environmental Protection Administration (SEPA) and the State Economic and Trade Committee (SETC) (*China Environmental News*, 20 January 2001). This kind of relaxation of regulation might give a sense of unfairness to other enterprises and discourage them from complying with the regulations. Second, there often emerge reports of non-compliance with the regulations. According to the reports of *China Environmental News*, sampling surveys in each provincial government have shown that 17 per cent in Zhenjiang, 30 per cent in Hebei, 40 per cent in Henan and nearly 50 per cent of enterprises in the major river basin of Shandong did not comply with emission standards or bans on production (*China Environmental News*, 2 May, 30 May, 11 July and 18 July 2001).

It should be noted that there is an institutional background of failures in enforcing environmental regulations. As economic reform has deepened, local governments have faced severe budget constraints, forcing them to dedicate themselves to raising operating funds from local enterprises to meet local economic and social demands. This is thought to be the political and economic base of localism, often called 'local protectionism' (*difang baohu zhuyi*), in China.[22] From this viewpoint, it can be said that local governments that receive economic benefits from local enterprises are reluctant to cooperate with environmental protection projects because they are afraid that the cost of environmental protection could worsen their budget constraints in the short term. This is one aspect of the explanation for the enforcement failure of environmental policy in contemporary China.

Local government, mainly at the level of district and city, fines the enterprises and imposes administrative punishment on the persons who will not comply with the regulations, but there is a ceiling on the fines, so they are not high enough to prevent violations. Although criminal penalties have been introduced in some environmental laws, there are very few cases where criminal penalties have been applied to pollution issues. SEPA has organized intensive inspections of environmental non-compliance with other ministries and bureaus since 2001. Mass media has cooperated with SEPA to report a lot of environmental non-compliance matters in newspapers, radio and TV. These intensive inspections can be considered a kind of top-down mobilization, like the National Inspection for Environmental Law Enforcement initiated in 1993 that used mass media campaigns to promote local environmental enforcement by building political and social pressures (*China Environmental News*, 26 September 2001). This means that intensive political pressure is needed to supplement the weak daily enforcement by environmental protection administrations.

Incentives to industrial firms for better environmental technologies

The Institute of Developing Economies, Japan External Trade Organization (JETRO) and Qinghua University in China jointly conducted a questionnaire survey on wastewater treatment facility installation at 100 industrial firms in Jiangsu Province in 2001.[23] Jiangsu Province is a province situated in the coastal region to the west of Shanghai, and it contains the lower reaches of the Huai River and the Lake Tai, both of which have been targeted as a major water pollution control watershed by the State Council since the late 1990s. This survey covered 38 chemical and pharmaceutical industries, 31 textile and dyeing industries, 16 food and beverage industries, and 15 paper-making industries, which are the major water polluter industries in Jiangsu Province. Major findings from this survey are as follow (see Tables 8.4 and 8.5):

1. 96 firms use biochemical wastewater treatment technology at Level II.
2. 62 firms did not need to replace manufacturing processes for installation of wastewater treatment and only two firms have replaced all manufacturing processes for wastewater treatment.
3. 72 firms point out that they are facing high operation costs, 43 firms point out that they are facing high energy consumption and 16 firms say they are facing unstable operation.
4. As for sources of treatment technology, 54 firms have introduced technology from planning and research institutes, 40 firms have introduced it from environmental equipment manufacturing enterprises and 24 firms have bought standard equipment on the domestic market. Only nine firms have developed facilities by themselves, and only nine firms have imported facilities from abroad.
5. 73.5 per cent of the total cost of treatment facilities was paid by the company itself.
6. Over a quarter of firms have not enjoyed any preferential treatment from the government for their investment in installation of wastewater treatment facilities.
7. 81 firms say new laws and standards are the major driving factor for installation of wastewater treatment facilities and 79 firms say enhancement

Table 8.4   Installation of industrial wastewater treatment facilities in 100 firms in Jiangsu Province (1)

Industries		Chemical	Dyeing	Foods	Pulp	100 firms
Number of firms		38	31	16	15	100
Number of facilities		45	36	17	15	113
Treatment technology level	Preliminary treatment	0	0	0	0	0
	Primary treatment	4	1	0	1	6
	Secondary treatment	38	35	17	14	104
	High-degree treatment	2	0	0	0	2
	After treatment	1	0	0	0	1
Reform of production	Wholly reformed	1	0	1	0	2
process by installation of	Partially reformed	16	11	3	6	36
facilities	No reform required	21	20	12	9	62
Problems of facilities	Running cost	30	20	9	13	72
	Energy consumption	18	14	7	4	43
	Instability of running	5	7	2	2	16
	Maintenance	4	5	5	0	14
	Treatment technology	6	5	1	1	13
	Treatment efficiency	4	3	2	2	11
	Operation	1	4	0	0	5
	Capacity of operators	0	3	0	0	4
	Frequency of trouble	0	1	0	0	1

Source: Compiled by the author based on data of 2001a 'Survey on Wastewater Treatment Facility Installation in Jiangsu, China'.

Table 8.5　Installation of industrial wastewater treatment facilities in 100 firms in Jiangsu Province (2)

Industries		Chemical	Dyeing	Foods	Pulp	100 firms
Total		38	31	16	15	100
Source of technology	Self-developed	5	3	0	1	9
	Planning institutes	21	17	10	6	54
	Environmental equipment manufacturers	18	13	3	6	40
	Purchasing of domestic standard equipment	9	6	5	4	24
	Import	1	0	3	5	9
	Others	0	3	1	0	4
Financing (%)	Environmental investment for basic infrastructure	13.1	1.6	0.6	3.8	3.0
	Environmental investment for planned projects	4.2	0.6	0.3	1.8	1.1
	Subsidy from pollution charges	12.3	6.9	0.9	5.9	4.0
	Bank loan	6.5	3.8	1.8	2.8	2.9
	Foreign funds	3.4	12.7	22.0	0.0	15.4
	Self-financing	59.4	74.0	74.5	85.7	73.5
	Others	1.2	0.4	0.0	0.0	0.2
Preferential treatment	Subsidy	16	10	6	7	39
	Exemption from pollution charge	11	6	9	4	30
	Low-interest loan	7	3	2	2	14
	Tax exemption	0	0	0	1	1
	Others	0	0	0	1	1
	None	11	15	6	5	37
Major factors for installation of facilities	New laws and standards	31	24	14	12	81
	Enhancement of regulatory enforcement	26	26	14	13	79
	Preferential treatment	9	6	4	2	21
	Complaints by residents	7	9	3	1	20
	Publicity by mass media	6	7	4	3	20
	State of business management	23	19	5	6	53
	Reform of enterprise system	11	13	4	2	30

Source: Compiled by the author based on data of 2001a 'Survey on Wastewater Treatment Facility Installation in Jiangsu, China'.

of enforcement is the major one, while only 20 firms say public pressure surrounding the firm and reports by mass media is the major driving factor.

These findings reveal that 100 industrial firms in Jiangsu Province (1) have installed wastewater treatment facilities since new and enhanced regulations for industrial pollution were adapted in late 1990s (see Table 8.4); and (2) have not always enjoyed preferential treatment from the government for wastewater treatment investment and have mostly utilized their own funds (see Table 8.5); and (3) have introduced end-of-pipe technology mostly from the domestic market and their level of in-house development is not so high. More than 10 per cent of the firms say the operation of their facility is not stable. That is to say, industrial firms in Jiangsu Province have introduced the 'second best available technology' under incomplete regulatory enforcement and weak public pressure.

The role of public pressure and participation, and their limitations

As mentioned earlier in this chapter, information disclosure and public participation have been some of the important approaches that complement regulatory enforcement of industrial pollution control in China since the 1990s; however, it should be noted that the anticipated role of the 'public' is supervision of polluters and local government under the basic environmental and social policy established by CCP and the central government. Furthermore, the 'public' in China is not allowed to raise any objection to basic policy, not only for environmental problems but also for other social issues, set by CCP and the central government.

For example, the two sets of regulations concerning the registration and management of NGOs stipulate in explicit terms that no non-governmental organization may oppose the basic principles of the Constitution (in particular, guidance by the Chinese Communist Party), harm national unity or security, harm particularly national unity, harm national interests, social interests, or the interests of other organizations and citizens, or carry out acts contrary to social and public morals (GFZ *et al.*, 1999, pp. 18 and 113).

As another example, it should be noted that limits on information disclosure related to environmental issues have been stipulated clearly in administrative regulations. What kinds of information are to be kept secret and classified have been listed in the Rules and Scope of State Secrecy in the Environmental Protection Project enacted in 1990 and revised in 1996. During these six years, environmental information has been more and more widely disclosed since formerly secret medium-term plans were exempted from secrecy by the rules revisions of 1996. On the other hand, detailed reports on the quality of the environment at the national and provincial levels are still secret, and even research reports and data on pollution incidents with serious impact and pollution diseases have become classified since 1996, indicating an increase in 'secrecy' (GHJF, 1993, pp. 292–3; GHZB, 1999, pp. 45–7). CLAPV, mentioned above, also faces these limitations on environmental information when they conduct surveys on victims of pollution incidents and disease.

Even under such political constraint, SEPA and local environmental protection bureaus need public pressure and participation to improve the weak enforcement of industrial pollution control. Also, some areas of government favour the introduction of a method of information disclosure and public participation because it could offer a low-cost approach, given the limited human, monetary and technical resources for monitoring all polluters and the administrators who are responsible for controlling polluters.

It cannot be denied that information disclosure and public participation initiated by the government and NGOs could enhance political and social pressure above the initial expectations. As seen in Figure 8.4, people's letters of complaint about environmental pollution to environmental protection bureaus and SEPA have increased since the late 1990s when the State Council issued the important decision enhancing pollution control all over the country, and this trend has continued. SEPA also focuses its inspections on serious

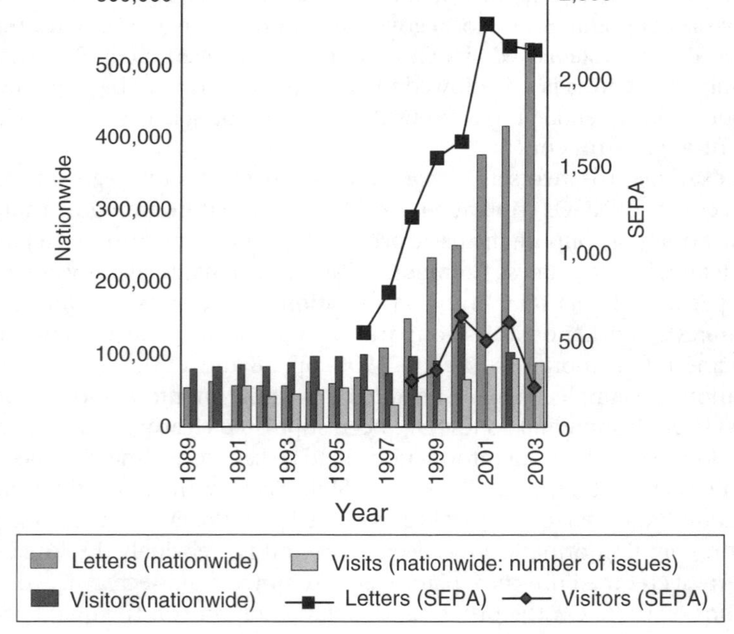

Figure 8.4   Complaints by people regarding environmental pollution problems
Note: The data on visitors in 2001 and 2003 are not available.
Sources: Compiled by the author, from *China Environmental Yearbook (ZHNBW 1990–2003)* and *Annual Report of China Environmental Statistics 2003 (GHZ, 2003)*.

environmental problems that attract great concern and the many requests for solutions from people, which are felt as high social pressure.

How such political and social pressures could exert influence upon the incentive of polluters to stop pollution should be examined. Dasgupta *et al.* (2000) discuss in their research paper how complaints from citizens could exert a strong influence upon governmental supervision and inspections of polluters, but that citizens' complaints did not exert direct influence upon polluters' behaviours. As mentioned in the previous section of this chapter, the questionnaire survey of 100 industrial firms in Jiangsu Province revealed that only 20 firms say public pressure surrounding the firm and reports by the mass media were major factors influencing their installation of wastewater treatment facilities. Also, some macro-statistical data tell us there could not be a direct connection between public pressure and pollution control investment. As seen in Figures 8.3, 8.4 and 8.5, there is an increase in industrial pollution control investment around the year 2000, although there also was an increase in people's letters of complaint about environmental pollution to the environmental administration. These findings imply that public pressure itself is still not strong enough to push polluters to take measures to stop pollution in China.

The Industrial Environmental Performance Disclosure Programme conducted in Jiangsu and other provinces and cities could be a possible means of exerting influence upon the incentive of firms through public pressure. Table 8.6 shows changes in the rating of industrial firms' environmental performance in three cities in Jiangsu Province. A prominent improvement can be seen in Yangzhou City, with an increase in blue-coloured firms from 68–85 per cent in one year, but indicators in the other two cities show no remarkable improvement. This programme has just begun, so we should continue to observe the impact of this information disclosure programme on environmental performance improvement over the long term.

Table 8.6   Changes in rating of environmental performance of industrial firms in three cities in Jiangsu Province

Rating Month/year	Zhenjiang			Yangzhou		Huai an	
	7/2000	6/2001	6/2002	6/2001	6/2002	6/2001	6/2002
Green	1 (1)	2 (2)	3 (2)	0 (0)	2 (2)	1 (2)	3 (6)
Blue	55 (61)	64 (61)	90 (68)	36 (68)	81 (85)	35 (76)	36 (75)
Yellow	21 (23)	24 (23)	26 (20)	12 (22)	7 (8)	4 (9)	5 (11)
Red	11 (12)	13 (12)	10 (8)	4 (8)	4 (4)	5 (11)	3 (6)
Black	3 (3)	2 (2)	3 (2)	1 (2)	1 (1)	1 (2)	1 (2)
Total	91 (100)	105 (100)	132 (100)	53 (100)	95 (100)	46 (100)	48 (100)

Note: Figures given in parentheses are percentages.
Source: Lu *et al.* (mimeo).

5.   Conclusion

China stepped towards building a multi-stakeholder governance system for implementing environmental policy and has enhanced industrial pollution control under this system since the 1990s. Furthermore, as it adopts new measures for information disclosure and the entrance of new actors such as NGOs, China has faced a restructuring of the multi-stakeholder cooperative relationship in environmental governance in recent years. This transformation of the governance system will result in more effective enforcement of industrial pollution control. However, we cannot draw a simple blueprint for future development of environmental governance in China, not only because the impact of the transformation is not clear yet, but also because the ruling party and government still tighten the screws of political liberty.

Under such a situation of environmental governance in China, it should be noted that there are some pilot projects to facilitate roundtable discussions among stakeholders to monitor industrial polluters' behaviour, jointly conducted by SEPA and the World Bank. Some environmental NGOs that are concerned with helping pollution victims, such as the Huai River Protector (Institute of Water Environment Ecosystem in the Huai River Basin),[24] also think that dialogues among stakeholders, monitoring of polluters by pollution victims and a broader spectrum of residents are inevitable and will be effective measures in China. Further examination is necessary on how to set up such a roundtable discussion among stakeholders for pollution control while reforming the existing governance system in China. Another possibility for enhancing enforcement of environmental policy is lawsuits by victims, which CLAPV is assisting. Recently, there has been a visible increase in awareness concerning the fight against environmental pollution not only among people in urban areas but also rural areas in coastal regions. But, as seen in this chapter, it will be a long and winding road to realize the environmental rights of people in China.

Notes

1. This chapter is revised from the paper presented at the 2003 Berlin Conference on the Human Dimension of Global Environmental Change, 'Governance for Industrial Transformation', held at the Environmental Policy Research Centre, Freie Universität Berlin, 5 December 2003.
2. The stipulation of Principle 10 is as follows. 'At the national level, each individual shall have appropriate access to information concerning the environment that is held by public authorities, including information on hazardous materials and activities in their communities, and the opportunity to participate in the decision-making process. States shall facilitate and encourage public awareness and participation by making information widely available. Effective access to judicial and administrative proceedings, including redress and remedy, shall be provided' (http://www.un.org/documents/ga/conf151/aconf15126-1annex1.htm).

3. For recent studies on environmental policy implementation, see Sinkule and Ortolano (1995), Ma and Ortolano (2000), Wang (2000), Economy (2004), etc.

4. Major cooperative research projects with Chinese counterparts in which the author has participated are: an environmental awareness survey in Beijing and Shanghai (FY1993–6); a study on environmental pollution disputes and environmental policy in China (FY2000); and a survey of industrial wastewater treatment facility installation in Jiangsu (FY2001). Each study included interviews and collection of materials. The first and last ones included an original questionnaire survey. For details of each study, see Nishihira *et al.* (1997), Wang *et al.* (2001) and Gu *et al.* (2002).

5. For example, refer to the lecture by Qu Geping, who was a top administrative leader of environmental policy in China, at China Central TV in 1980 (Qu, 1984, pp. 295–8).

6. NEPA was reformed as SEPA (State Environmental Protection Administration) in 1998.

7. For details on the history of the environmental administrative organization in the State Council of China, see Kataoka (1997), Cheng and Piao (1994, pp 43–6) and Qu (1997, pp. 263–72).

8. The first environmental protection administrative body at the state level was the State Council Leading Group on Environmental Protection set up in 1974. The group was abolished and the Environmental Protection Bureau was newly set up under the Ministry of Urban and Rural Environmental Protection in 1982. It was the predecessor of NEPA (Chen and Piao, 1994, pp. 43–6; Qu. 1997, Intro., pp. 14–15).

9. Besides *China Environment News*, administrative information on environmental policy has been revealed in the Bulletin of China Environmental Situation and *China Environmental Yearbook* since 1990 (ZHNBW, 1990–2002).

10. Although it may be true that the Chinese government has participated in the international environmental community with awareness of its own important role in fighting against environmental deterioration in both the domestic and international contexts, it was also revealed that there was a diplomatic necessity behind its participation. The document issued by the Ministry of Foreign Affairs in March 1992 noted the importance of UNCED in freeing China from the economic isolation forced on it by Western countries after 4 June 1989 (GHWM, 1995, pp. 520–1), when the People's Liberation Army suppressed the pro-democracy movement led by students and citizens in Beijing.

11. The Century Walk was said to be initiated by Qu Geping (Qu 1997, Intro., pp. 20–1; Zhe, 1998, p.145).

12. Since 1994, the Ministry of Agriculture, Ministry of Water Resources, Central Youth Organization of Communists, eight mass media organizations and China Foundation for Environmental Protection have participated in the organizing committee of the campaign (ZHNBW, 1995, pp. 21–3).

13. *China Youth Daily*, 4 August 1994, and *People's Daily*, 13 August 1994.

14. The policy process on water pollution problems in the Huai River is mainly based on interviews and materials collected by the author in China during the period from 1997–9.

15. See Note 2.

16. The role of the stock market was not thought to be a major factor due to its immaturity in China (Wang *et al.*, 2002, p. 185).

17. In villages, the method is different from in the cities, with the holding of round-table discussions with the participation of local government, local assembly members, firms and villagers (from an interview with Professor Lu, Nanjing University, 2003).

18. According to Zhao (1999), the number of non-governmental environmental protection organizations in Beijing is as follows: 100 or more formal social associations, 15 students' associations and six 'grassroots' organizations, classified as such in this chapter (of which three are informal organizations of volunteers).
19. CLAPV is the popular English name for the centre, and it reflects the centre's mission. In Chinese, the centre's registered name is *Huanjing Ziyuan Fa Yanjiu He Fuwu Zhongxin* (Centre for Research and Service of Environmental and Natural Resource Law).
20. Refer to CLAPV (2002) and the website of CLAPV (http://www.clapv.org/, accessed on 19 August 2005).
21. CLAPV has also conducted workshops to promote communication and discussion for better environmental law enforcement with Chinese stakeholders and foreign counterparts. CLAPV and the Japan Environmental Council (JEC) held a China-Japan Workshop on Environmental Disputes Settlement in Beijing in September 2001 and in Kumamoto in March 2004. Besides the workshops with JEC, CLAPV held a Workshop on Difficulties in Environmental Lawsuits in the Western Part of China, in Xian in October 2002, and an International Symposium on Lawmaking of Compensation for Environmental Damage in Beijing in August 2004.
22. Oi (1992) calls such localism 'local state corporatism'.
23. A portion of the results of this survey is summarized in Gu *et al.* (2002).
24. The Huai River Protector is a grassroots environmental NGO, which was registered as a non-governmental environmental organization at the Bureau of Civil Affairs, Shenqiu County, Zhoukou City, Henan Province, China, on 23 October 2003. It focuses on cleaning up water in the Huai River to promote the building of a public participation mechanism, as well as to defend public environmental rights for sustainable development at the grassroots level in the whole river basin. See the website at: http://www.lwlsw.com/hhwsw/

References

Centre for Legal Assistance to Pollution Victims (CLAPV) (2002) *Zhongguo zhenghua daxue huanjing ziyuanfa yanjiu he fuwu zhongxin nianjian* 2002 (*Yearbook of CLAPV 2002*) (in Chinese), China University of Political Science and Law.

Chen Hanguang and Piao Guangzhu (eds) (1994) *Huanjing fa jichu* (*Basic Environmental Law*) (in Chinese), Beijing, Zhongguo shehui kexue chubanshe.

Chen Hanguang and Piao Guangzhu (eds) (1997) *Huanjing zhifa jichu* (Basic Environmental Law Enforcement) (in Chinese), Beijing, Falv chubanshe.

Dasgupta, Sumita, Benoit Laplante, Nlandu Mamingi and Hua Wang (2000) 'Industrial Environmental Performance in China', Policy Research Working Paper 2285, Washington, DC, World Bank, Development Research Group, Infrastructure and Environment.

Economy, Elizabeth (2004) *The River Runs Black: The Environmental Challenge to China's Future*, Ithaca and London, Cornell University Press.

Gu, Shuhua and Xiliang Zhang, Bingxiang Ren and Baoling Guo (2002) 'A Survey of Wastewater Treatment Facility Installation in China: Opportunities for International Cooperation of Technology Transfer,' IDEAS Machinery Industry Study Report No. 2, Chiba, Japan, Institute of Developing Economies, Japan External Trade Organization.

Guowuyuan Fazhiban Zhengfa-si and Mingzheng-bu Minjian-zuzhi Guanli-ju (GFZ *et al.*) (eds) (1999), Shehui-tuanti Dengji Guanli Tiaoli, Minban-feiqiye-danwei Dengji Guanli Zanxing-tiaoli, Shiyi (Commentaries of Regulation of Registration and Management on Social Associations and Provisional Regulation of Registration and Management on Non-govenmental and Non-Industrial Enterprises) Beijing, Zhongguo Shehui Chubanshe, (in Chinese).

Guojia Huanjingbaohu Ju Faguisi (GHJF) (ed.) (1993) *Huanjing baohu fagui huibian* (Collected Laws and Regulations of Environmental Protection) (in Chinese), Beijing, Zhongguo huanjing kexue chubanshe.

Guowuyuan Huanjingbaohu Weiyuanhui Bangongshi (GHWB) (ed.) (1988) *Guowuyuan Huanjing Baohu Weiyuanhui Wenjian Huibian* (*Archives of the State Council Committee on Environmental Protection*) (in Chinese), Beijing, Zhongguo huanjing kexue chubanshe.

Guowuyuan Huanjingbaohu Weiyuanhui Mishuchu (GHWM) (ed.) (1995) *Guowuyuan Huanjing Baohu Weiyuanhui Wenjian Huibian II* (*Archives of the State Council Committee on Environmental Protection, Vol. 2*) (in Chinese), Beijing, Zhongguo huanjing kexue chubanshe.

Guojia Huanjingbaohu Zongju (GHZ) (ed.) (2003) *Zhongguo Hyanjing Tongji Nianbao (Annual Report of China Environmental Statistics)*, *2003*, Beijing, Zhonguo huanjing kexue chubanshe. (in Chinese)

Guojia Huanjubaohu Zongju Bangongting (GHZB) (ed.) (1999) *Huanjing baohu fagui huibian (Collected Laws and Regulations of Environmental Protection)* (in Chinese), Beijing, Zhongguo huanjing kexue chubanshe.

Guojia Huanjingbaohu Zongju Bangongting (GHZB) (ed.) (2000) *Huanjing Baohu Wenjian Xuanbian* (*Document Collections of Environmental Protection*) (1999) (in Chinese), Beijing, Zhongguo huanjing kexue chubanshe.

Kataoka, Naoki (1997) *Chugoku kankyo-osen bouchi-hou no kenkyu (A Study on Environmental Pollution Regulations in China)*, (in Japanese), Tokyo, Seibundo.

Kobayashi, Masayuki (1999) 'Chugoku no kenpo seido' ('Constitution in China'), in Yasuki Omura and Masayuki Kobayashi (eds), *Higashi ajia no kenpo seido* (*Constitutions in East Asia*) (in Japanese), Tokyo, Ajia keizai kenkyusho (Institute of Developing Economies).

Leading Group for China's Agenda 21 (LGCA) (1994) *China's Agenda 21: White Paper on China's Population, Environment and Development in the 21st Century*, Beijing: China Environmental Science Press.

Lu Genfa, Wang Yuan, Qian Yu, and Hua Jun, 'Qieye Huanjing Xingwei Xingxi Gongkai Zai Zhongguo Huanjing Guanli Zhongde Yingyong' (A Practice of Industrial Environmental Performance Disclosure in Chinese Environmental Management) mimeo (unpublished).

Ma, Xiaoying and Leonard Ortolano (2000) *Environmental Regulation in China: Institutions, Enforcement, and Compliance*, Maryland, Rowan & Littlefield.

National Environmental Protection Agency (NEPA) (1996) *Disici quanguo huanjingbaohu huiyi wenjian huibian* (*Archives of the Fourth National Conference on Environmental Protection*) (in Chinese), Beijing, Zhongguo huanjing kexue chubanshe.

Nishihira, Shigeki, Reeitsu Kojima, Hideo Okamoto and Shigeaki Fujisaki (eds) (1997) *Environmental Awareness in Developing Countries: The Cases of China and Thailand*, Tokyo, Institute of Developing Economies.

Oi, Jean C. (1992) 'Fiscal Reform and the Economic Foundations of Local State Corporatism in China', *World Politics*, 45, October.

Otsuka, Kenji (2002a) 'China: Social Restructuring and the Emergence of NGOs', in Shinichi Shigetomi (ed.), *The State and NGOs: Perspective from Asia*, Singapore, Institute of Southeast Asian Studies.

Otsuka, Kenji (2002b) 'Chugoku: shuto pekin no kankyo mondai ga tou gyosei-simin kankei' ('China: Government and Citizens in Environmental Issues in the Capital City, Beijing'), *Ajiken World Trend* (Chiba, Japan) (in Japanese), 76, 8–11 January.

Qu Geping (1984) *Zhongguo huanjin-wenti ji duice* (*China's Environmental Problems and its Countermeasures*) (in Chinese), Beijing, Zhongguo huanjing kexue chubanshe.

Qu Geping (1997) *Women xuyao yichang biange* (*We Need a Change*) (in Chinese), Changchun, Jilin renmin chubanshe.

Sinkule, Barbara J. and Leonard Ortolano (1995) *Implementing Environmental Policy in China*, London, Praeger.

State Environmental Protection Administration (SEPA) (2001) 'Official Bulletin of National Survey on Industries Concerning Environmental Protection'.

Wang Canfa, Xu Kezhu, Hu Jing, Liu Min, Tadayoshi Terao and Kenji Otsuka (2001) *Studies on Environmental Pollution Disputes: Cases from Mainland China and Taiwan (Joint Research Programme Series No.128)*, Chiba, Japan, Institute of Developing Economies, Japan External Trade Organization.

Wang Hua (2000) 'Pollution Charges, Community Pressure, and Abatement Cost of Industrial Pollution in China', Policy Research Working Paper 2337, Washington, DC, World Bank, Development Research Group, Infrastructure and Environment.

Wang Hua, Cao Dong, Wang Jinnan, Lu Genfa, *et al.* (2002) *Huanjing xinxi gongkai linian yu shijian* (*Environmental Information Disclosure: Concept and Practice*) (in Chinese) Beijing, Zhongguo huanjing kexue chubanshe.

World Bank (2000) *Greening Industry: New Roles for Communities, Markets and Governments*, New York, Oxford/World Bank.

Xie Jingtao (ed.) (1998) *Zhongguo difang zhengfu tizhi gailun* (*Overview of Local Government System in China*) (in Chinese), Beijing, Zhongguo guangbo dianshi chubanshe.

Xu Kezhu (2004) 'You-xibu-huanjing-weiquan-susong-yinfade-ruogan-sikao' ('A Prospect Caused by Experiences in Environmental Lawsuits in Western China'), (in Chinese), a report presented at the Second China–Japan Workshop on Environmental Disputes Settlement, held in Kumamoto, Japan, 20 March.

Zhao Xiumei (1999) *Beijing-shi-huanjing-baohu-gongzhong-canyu-jizhi-yanjiu* (*A Study on Public Participation Mechanisms in Environmental Protection in Beijing*), Qinghua-daxue-NGO-yanjiu-zhongxin-baogao (Report of NGO Research Centre), Beijing, Qinghua University NGO Research Centre.

Zhe Fu (1998) *Zhongguo dangan* (*China Documents*) (first volume), Beijing, Guangming ribao chubanshe.

Zhongguo Huanjing Nianjian Bianji Weiyuanhui (ZHNBW) (ed.) (1990–3) *Zhongguo Huanjing Nianjian* (*China Environmental Yearbook*) (in Chinese), Beijing: Zhongguo huanjing kexue chubanshe.

Zhongguo Huanjing Bianji Weiyuanhui (ZHNBW) (ed.) (1994–2002)
Zhongguo Huanjing Nianjian (*China Environmental Yearbook*) (in Chinese), Beijing, Zhongguo huanjing nianqianshe.

Zhongguo Huanjingbaohu Xingzheng 20 Nian Bianweihui (ZHX) (ed.) (1994) *Zhongguo Huanjing Baohu Xingzheng 20 Nian* (*China Environmental Administration 20 Years*) (in Chinese), Beijing, Zhongguo huanjing kexue chubanshe.

9

Democratization, Decentralization and Environmental Policy in Taiwan: Political Economy of Environmental Policy Formation and Implementation

Tadayoshi Terao

Introduction

Taiwan experienced rapid industrialization starting in the 1960s and suffered severe environmental problems, such as aggravation of industrial pollution, degradation of the living environment and degradation of natural resources, in the process. The countermeasures for environmental pollution and the deployment of environmental policy were not at all sufficient to counteract environmental degradation.

Although a legal system for water and air pollution control had been enacted since the middle of the 1970s, the levels of pollution emission regulation were loose and establishment of an administrative organization to support implementation was not fully realized. It may be said that progress towards a full-scale environmental administration in Taiwan started the second half of the 1980s.

Recognition of the seriousness of the environmental pollution problem by the government, which promoted rapid industrialization and suppressed political freedom, was delayed because the people's dissatisfaction with the pollution problem also continued to be suppressed. When political liberalization and democratization advanced starting in the early 1980s, the government, private enterprises and citizens did not have enough knowledge of effective measures or the experience to tackle environmental pollution. As a result, many disputes over pollution continuously occurred and proper dispute resolution could not be achieved, which led to social confusion.

With regard to previous studies on the environmental issues and environmental policy of Taiwan, some studies in the early 1990s focused on environmental problems generated in the shift from the authoritarian regime to democracy through which economic development was promoted and on the countermeasures for the environmental problems from the viewpoint of political economy. In this chapter, we will present the changing background of environmental policy from the second half of the 1990s, as well as the outcome that was achieved in Taiwan after the establishment of an environmental policy administration in the latter half of the 1980s. Furthermore, we will examine why and how we need to change our basic ideas on 'development and environment' in the process of democratization and decentralization, by presenting a variety of experiences from local environmental politics in Taiwan in the late 1990s and the early 2000s. As a whole, we will attempt to evaluate not only the current situation of environmental policy in Taiwan, but also the basic ideas presented in the analysis of the early 1990s.

In Section 1, we will explain the historical development of environmental policy administration in Taiwan and its political and economic background. In Section 2, trends of the environmental quality indexes, including air and water, will be shown and analysed. In Section 3, we will investigate the effectiveness of pollution regulation implementation, by using statistical data from local government inspections of pollutant sources. In Section 4, as an important actor in environmental policy implementation, local governments and their political background will be analysed, using as examples three county governments. We can find a large variety of attitudes towards environmental protection among local governments. Although democratization and decentralization changed the environmental policies in Taiwan, the relationship between the changes in the direction of environmental policy and changes in political conditions is not a simple one.

1.   Emergence of environmental issues, and deployment of environmental administration

The economic growth of Taiwan after the Second World War was remarkable, and during the nearly 40-year period from 1952–91, real GDP growth attained 8.7 per cent on average. Although there was fluctuation in the short term, the annual averages of real GDP growth rates were close to 10 per cent during the 1970s and the 1960s. Per capita GDP exceeded US $10,000 in 1991.

The load placed on the natural environment in Taiwan quickly expanded as a result of such rapid economic growth and industrialization. Without proper countermeasures and environmental policies, the damage caused by the expanding load could not be prevented. However, development of an effective environmental administration was long overdue, given the high speed of economic growth and rapid expansion of the load on the environment.[1]

The KMT (Kuomintang: Chinese Nationalist Party) administration of Taiwan, which justified oppression of political freedom by promotion of economic development, gave priority to economic growth for many long years and was unable to recognize the seriousness of the industrial pollution that accompanied rapid industrialization. Therefore, provision of measures to deal with the pollution problem was delayed, and the industrial pollution situation became increasingly serious.

People's dissatisfaction and protests against environmental pollution and large-scale industrial development projects were suppressed, and the exposure of the problems and provision of countermeasures delayed sharply. As a result, many large-scale development projects that placed a heavy load on the environment and neighbouring communities were carried out without proper environmental impact assessments and countermeasures.

In parallel with the progress of political liberalization and democratization from the first half of the 1980s, anti-pollution movements and environmental protection movements occurred in many different regions of Taiwan. Those movements were the impetus towards promotion of environmental policy administration and institutions and pressured private enterprises to undertake environmental pollution prevention. Improvement of the environment began only after the government enhanced the system and organization of the environmental policy administration.

Citizens' movements asking for political freedom and democratization and protest movements against environmental pollution were staged in parallel and boosted each other. The social pressures they created helped to promote the development of pollution control measures by the government and private enterprises. The environmental protection movement energized the movements for political freedom and democratization, and political freedom and democratization, which were realized gradually, enabled citizens to express further dissatisfaction with environmental pollution problems, thus stimulating the development of an environmental administration.[2]

In 1971, the Department of Health became independent from the Ministry of the Interior, within the Executive Yuan (the central government administration). Before this upgrade, even the public health and sanitation administration, which could be considered an early stage of environmental administration, was not an independent organization within the central government.

At the local government level, the Department of Health was established by the Taiwan Provincial Government at that time. In the Taipei Municipal Government, the Environmental Health Department was established in 1968. In each county and municipal government under the Taiwan Provincial Government, sections in charge of environmental sanitation were established within each of the public health bureaus in 1962.

The various laws and orders to regulate environmental pollution were enacted after the establishment of the Department of Health in the central government in 1971. The Water Pollution Prevention Law was proclaimed in 1974, and the Air Pollution Law was proclaimed in 1975. Furthermore, the Waste Management Law was proclaimed in 1974. However, implementation was not effective as there was no improvement in the actual enforcement of regulations or the standards for pollutant emission and concentration.

The Division of Environmental Protection was upgraded to the Environmental Protection Bureau in the Department of Health of the central government in 1982, and it had jurisdiction over the air pollution control and water pollution control, which were taken away from other sections in the central government. Also in the Taiwan Provincial Government, various kinds of anti-pollution measures that had been dispersed among sections were centralized, and at the same time the Department of Environmental Protection was established in the Taiwan Provincial Government. Enforcement provisions for each regulation and law were also established at last, and a substantial pollution regulation policy and environment policy took shape at that time.

The Environmental Protection Administration (EPA) was established as an independent administrative organization in charge of environmental protection in the central government in August 1987. At this time, the pollution problems and pollution incidents were surfacing in various parts of Taiwan, in tandem with the gradual progress of political liberalization and democratization that had been ongoing since the early 1980s. By this time, environmental pollution had already become a serious social problem. The spontaneous protests against the existing industrial pollution and the new development projects by local residents and communities were called 'self-relief' movements. In the self-relief movement against pollution, victims of pollution and other concerned persons formed temporary organizations called 'self-relief associations'. They travelled to polluting factories, development project offices, local and central governments, etc., and staged direct protests against them. Factories discharging pollutants were often forced to suspend their operations due to the collective use of force by such anti-pollution self-relief organizations. Although self-relief organizations involved in the anti-pollution and anti-development movement had been witnessed since the first half of 1980s, it was during the time of political liberalization in the second half of the 1980s, especially around the time martial law was lifted in July 1987, that those movements came to occur frequently in Taiwan.[3]

There was no large systematic change in the central government's environmental policy administration organization after the establishment of the Environmental Protection Administration in 1987. Under the Democratic Progressive Party (DPP) administration, which took office in 2000, in order to increase the independence of the Environmental Protection Administration further and also to concentrate the administrative authorities related to environmental policy, which was dispersed among other sections of the central government, the establishment of the Ministry of Environment and Resources was discussed in the central government.

With regard to legislation after the establishment of the EPA, the Pollution Dispute Resolution Law was enacted in 1992, and the Environmental Impact Assessment Law was enacted in 1994. These laws involve fields other than regulation of individual pollutant media, such as the atmosphere, water and solid waste. The existing laws and regulations that regulate water pollution, air pollution, and waste disposal, etc., were also reformed several times after the establishment of the EPA. Although there were extensive arguments from the end of the 1980s to the early 1990s concerning the Basic Environmental Law, which was to embody the government's overall plan for environment policy, no agreement on a final version was reached during that time. Even though no basic law for environmental policy was enacted throughout the 1990s, maintenance and revision were briskly performed on individual environmental laws. After the launch of the DPP administration in 2000, establishment of the basic law was attempted again, and the Basic Environmental Law was finally enacted in December 2002. In the Basic Environmental Law, the 'citizen lawsuit' is incorporated. Although the citizen lawsuit was already incorporated in the newest revision of the Air Pollution Prevention Law, the Water Pollution Prevention law and the Waste Management Law, it was specified anew by the basic law. However, heretofore there have been no substantial lawsuits that have utilized the 'citizen lawsuit'.[4]

2.   The situation of environmental quality

It may be said that substantial improvement of environmental quality can be observed as a result of development of the environmental administration systems mentioned above. In particular, the air quality situation has improved sharply, at least to the extent that a related statistical index is examined. This is interpreted to mean that a contribution has been made to the improvement of the air quality index through enforcement of measures against pollutant discharge such as SOx emission from industries, development of the mass transit infrastructure in urban areas that reduced the number of automobiles and measures to limit automobiles and motorcycles.

The state of air pollution monitored in Taiwan is displayed in Figure 9.1 in terms of PSI (Pollutants Standard Index), an index of air pollution.

Looking at the general air pollution situation in Taiwan as expressed by the PSI index over time, continuous improvement in air pollution is visible since the second half of the 1980s. The 'hazardous' rating, which is applied to the most serious air pollution situation, and the 'very unhealthy' rating are hardly observed after 1991. The frequency of the 'unhealthy' rating fell after 1992, and the frequency of the 'good' rating, which is the rating with the least contamination, rose continuously. Therefore, the air quality situation as indicated by the PSI index was improving continuously in Taiwan.

With regard to water quality as indicated by the measurement of pollution in major rivers, significant improvement has not yet been generally observed (Figure 9.2). At each monitoring point along major rivers in Taiwan, measurement results for each of the pollutants, BOD_5, DO (dissolved oxygen), SS (suspended substance) and ammonia, are indexed and classified into four

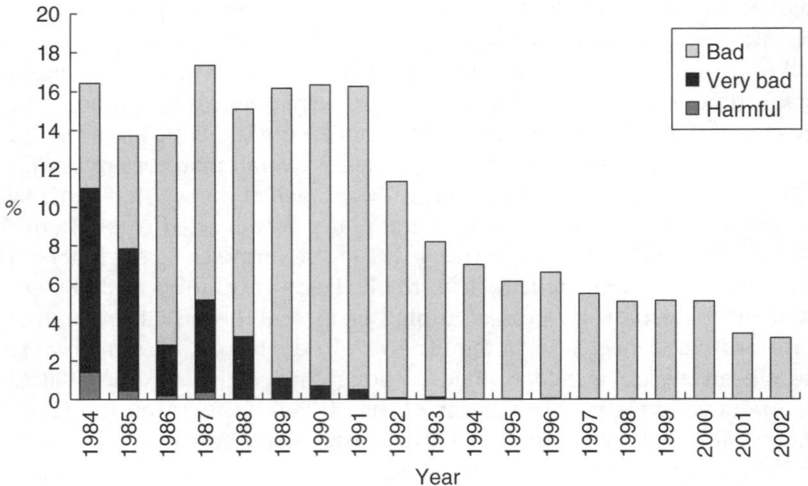

Figure 9.1 Air quality in Taiwan area: 1984–2002 (indexed by PSI)
Sources: Based on data from Environmental Protection Administration, 'Environmental Protection White Paper', various issues; Environmental Protection Administration, Executive Yuan.

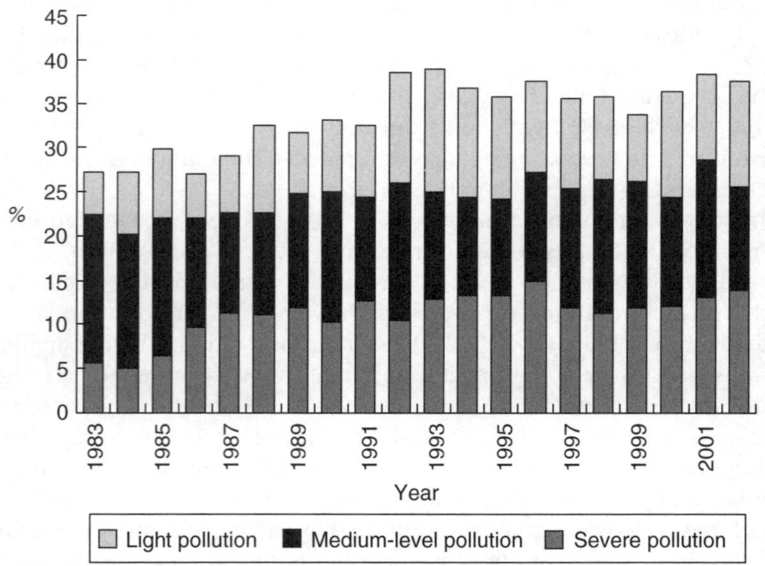

Figure 9.2 Water quality status of major rivers in Taiwan area (length, per cent)
Sources: Based on data from Environmental Protection Administration, 'Environmental Protection White Paper', various issues; Environmental Protection Administration, Executive Yuan.

categories which represent the degree of the contamination. The percentages shown in Figure 9.2 were calculated in terms of the length of major rivers based on the degree of contamination mentioned above. Following a quick increase in 1986 and 1987 in the contamination rate of portions of major rivers classified in the most serious category of 'severe contamination', the contamination rate shows a tendency towards aggravation in the long run or levelling off. Summing up the trends in the categories of 'severe', 'medium' and 'slight' contamination, until 1992 or 1993 a tendency towards aggravation is seen, and after that, no improvement is observed at all.[5]

3.   Effectiveness of the implementation of pollution discharge regulations by the environmental policy administration

It is difficult to quantitatively analyse the effectiveness of environmental pollution regulations implemented by the government. Although data is limited, we can examine the validity of the pollution emission regulations, including discharge from fixed-point sources such as factories of manufacturing industry, waste disposal sites and automobiles and motorcycles, of which local governments are mainly in charge.

Before analysing the overall tendencies using statistical materials, the structure of government administration for pollution discharge regulation in Taiwan will be explained briefly. In accordance with the regulatory laws for each pollutant media, such as air, water, solid waste and noise, environmental protection departments of local governments go to fix the sources of discharge, such as factories, and implement investigation. The EPA of the central government also has the authority to conduct investigations, as does the Bureau of Environmental Inspection, which maintains three branch offices, in northern, central and southern Taiwan. Each branch office has inspection teams. However, even if the inspection teams belonging to the Bureau of Environmental Inspection of the EPA investigate sources of pollution discharge and find actual violations, this organization does not have the authority to take direct action against an offender. Violations are dealt with through the proper staff in the local government. The on-site inspection by the central government's local branches can only complement inspections by the local governments, and the number of their inspections is limited compared to those of the local governments. It may be assumed that on-site inspections are mainly performed by local government.[6]

The annual number of inspections by the local governments for each type of pollution, such as air (fixed source discharge and traffic), noise, water and solid waste is shown in Part A of Table 9.1. Also, the percentage of inspections where inspection resulted in fine payment and the average amount of fine payment (NT$) per times of fine are shown in Parts B and C of Table 9.1. Water contamination mainly refers to manufacturing industries and animal

Table 9.1   Pollution source inspection and fine collection by local government

	1988	1989	1990	1991	1992	1993	1994	1995	1996	1997	1998	1999	2000	2001	2002
A. Number of pollution source inspections by local administration															
Air (stationary source)	20,207	26,227	43,587	41,396	46,016	51,836	42,040	40,268	58,492	61,247	65,605	61,552	67,460	68,171	68,784
Air (mobile source)	198,111	324,347	457,976	466,911	455,222	355,498	472,349	322,289	1,479,435	1,186,879	1,092,348	826,153	1,592,239	1,064,415	1,165,968
Noise	10,804	16,066	17,309	18,573	26,496	25,612	27,922	28,846	28,355	28,933	27,788	29,876	30,665	25,707	31,528
Water	12,541	24,443	44,161	43,861	42,678	32,838	29,332	27,470	27,686	28,807	39,319	42,149	44,383	29,055	29,598
Waste	96,714	128,750	133,190	121,823	115,623	128,856	157,003	208,777	202,167	182,631	170,134	198,860	194,110	228,260	242,063
Total	338,377	519,833	696,223	692,564	686,035	594,640	728,646	627,650	1,796,135	1,488,497	1,395,194	1,158,590	1,928,859	1,415,608	1,537,941
B. Violation (fine payment) rate (per cent)															
Air (stationary source)	14.65	18.26	13.11	13.96	10.63	6.90	7.35	6.50	3.74	3.88	4.61	4.42	5.07	3.94	4.29
Air (mobile source)	19.15	16.95	15.76	17.05	14.75	13.08	10.22	8.71	3.19	2.37	2.85	2.72	2.60	5.30	4.06
Noise	14.13	8.22	1.96	1.93	2.69	1.95	2.87	2.54	2.09	3.22	2.22	1.33	1.73	1.05	1.25
Water	50.39	48.76	31.25	27.17	25.75	15.53	15.52	19.01	19.22	18.90	17.09	11.12	8.90	11.04	11.17
Waste	73.98	85.31	59.71	60.11	55.88	54.82	53.33	52.63	47.70	44.40	31.89	30.03	27.76	17.82	23.25
Whole	35.55	35.17	24.64	24.68	21.62	21.24	19.27	23.35	8.45	7.92	6.86	7.77	5.35	7.30	7.17
C. Average fine payment per violation (NT$)															
Air (stationary source)	14,755	15,591	14,903	18,965	42,111	51,729	69,680	83,636	76,219	99,888	60,942	80,697	66,331	48,991	38,883
Air (mobile source)	2,124	1,810	1,552	1,878	2,333	2,690	2,861	2,733	2,491	3,369	2,446	1,852	1,741	1,741	2,150
Noise	1,676	1,554	5,182	5,601	4,948	6,441	4,763	6,278	8,465	5,575	9,911	8,859	4,638	9,351	9,219
Water	13,409	8,093	7,025	18,139	23,992	50,358	51,054	47,265	48,537	52,024	42,138	49,618	61,549	71,641	67,411
Waste	756	673	915	1,613	2,271	2,084	1,728	1,462	1,230	1,633	2,509	2,434	2,531	3,589	2,832
Whole	2,208	1,896	2,149	3,484	5,235	5,679	5,228	11,252	4,392	6,386	7,166	7,143	6,602	5,888	5,462

Sources: Calculated from Environmental Protection Administration, 'Yearbook of Environmental Statistics'; Environmental Protection Administration, Executive Yuan, various issues.

husbandry (hog raising), and 'air (stationary source)' mainly refers to manufacturing industries, while 'mobile source of air' refers to automobiles and motorcycles.

The fine payment rate (fine collection rate) reached approximately 36 per cent in total in 1988; in waste control the figure was 74 per cent and in water contamination the figure was approximately 50 per cent. When the local governments conducted on-site inspections at that time, rather than confirming compliance, it turned out that violations were discovered in many cases, resulting in collection of fines. Although regulations on pollution existed, it may be said that pollution regulation in the early stage had a limited result.

In general, a variety of factors may be considered as the cause of the ineffectiveness: (1) the existence of the regulations was not primarily recognized by the polluters; (2) the emission standards were too severe to be met technologically and financially at that time; (3) the polluters may have considered the probability of being inspected to be very low; and (4) the level of the fines (surcharge) was low compared to the cost to preventing pollution.

As already seen, the laws for regulating water pollution and air pollution were enacted in the first half of the 1970s, but detailed specific enforcement regulations were not enacted, so the effectiveness of these regulations was questionable. After the Bureau of Environmental Protection was established in the Department of Health of the central government in 1982, detailed enforcement regulations were enacted at last. However, organizations to enforce the laws were not fully established by each local government at that time. Moreover, the fines levied on violators were too low in most cases, and so did not provide enough incentive to private enterprises to reduce pollution. 'Paying the fine and continuing to discharge pollution' was far cheaper than paying the pollution prevention cost, and the level of regulation was also too loose.

After the EPA was established in 1987, the existing laws regulating each pollutant media were revised, and regulations became severer. The levels of fines were raised drastically and, at the same time, expansion of the local governments' environmental protection sections, which were to execute pollution regulations, proceeded quickly.

In 1988, regulatory standards were not necessarily so severe as to make them technically impossible. However, the regulations were not appropriately recognized by the parties discharging pollution. Moreover, the level of the fines was still not high enough. It was less expensive for polluters to pay fines than to install pollution reduction equipment even at large-scale factories where the probability of being inspected was comparatively high. If there had been no social pressure from NGOs or anti-pollution movements, which sought to help local residents and monitor the pollution discharge, private companies would not have had sufficient cause to reduce pollution emission.

Also, in the case of air pollution emitted from fixed point sources, as of 1988 fines were paid in about 15 per cent of the cases inspected. As already seen, the rate of violation had reached 74 per cent and 50 per cent in solid waste pollution and water contamination, respectively, and in the inspections by the local governments at this time, exposure of violations was the main goal, especially in case of inspections of waste and water pollution.

As for number of the inspections by the local governments, the number of inspections of water pollution and air pollution (fixed source discharge) was increasing rapidly in 1990. Although the number levelled off after that, it increased sharply again in 1996, and generally displayed a consistent increase thereafter.

The inspection and control of waste contamination increased sharply between 1989 and 1995. The probability that violators would be exposed rose, together with the increase in the number of inspections by the environmental administration departments, and this most likely caused the enterprises to reduce their pollution discharge to meet the regulatory standards.

Looking at fluctuations in the rate of fine payment, the fine payment rate for water contamination exceeded 50 per cent in 1988, fell to 25 per cent in 1990 and 1991, and then dropped sharply again in 1999 to around 10 per cent, which was maintained after that. The fine payment rate for air pollution (fixed source) was approximately 15–18 per cent in 1988 and 1989 but started to fall in 1990, fell sharply in 1993 and fluctuated within a 5 per cent range from 1996 onward. In case of waste pollution inspection, rate of fine payment also fell markedly. It peaked at 85.3 per cent in 1989 and fell to less than 50 per cent in the first half of the 1990s. It declined further in the late 1990s, to reach approximately 30 per cent in 1998, and in 2000 it dropped further to a low of 17.8 per cent.

On the other hand, the average amount of fine payments was increasing sharply, reflecting the rising maximum amount of fines levied by the government. In case of water pollution, for example, the increases in the average fine payment in 1991, 1992 and 1993 were remarkable. Following the revision of the Water Pollution Prevention Law in 1991, the maximum limit of fines for violation of emission standards by factories was actually raised from NT$60,000 to NT$600,000, an increase of ten times (according to the *Commercial Times*, 20 April 1991). Average fine amounts rose from NT$7,000 in 1990 to NT$50,000 or more by 1993. Corresponding to this rise in the average fine amount, the rate of fine payment fell from 31.3 per cent in 1990 to 15.5 per cent in 1993. The number of inspections by the local governments concerning water pollution dwindled at the same time. Although no causal relationship can be concluded from this data alone, the increase in the average fine amount is considered to be an important factor in the reduction of the rate of fine payment by violators of water pollution standards. Independent quantitative verification would be required to determine whether an average fine of NT$50,000 per violation was a sufficient amount of money to cause major private enterprises at this time to reduce pollution.

Regarding the fine paid for each air pollution (fixed source) violation, the amount increased rapidly, from NT$19,000 in 1991, to NT$42,000 in 1992, NT$52,000 in 1993, NT$70,000 in 1994 and NT$84,000 in 1995. The rate of fine payment by violators fell from 13.1 per cent in 1991 to 6.9 per cent in 1992 and 10.6 per cent in 1993, reflecting the increase. In the case of waste pollution, the rise in the fine amount per violation is not so sharp as in the cases of air pollution (fixed source) and water pollution. However, it is possible that in the second half of the 1990s, especially from 1998, the rapid decline in the rate of fine payment is related to the fact that there was a large increase in the fine amount per violation.

If decline in the rate of fine payment could be considered as an effect of regulation by the administration, then raising the maximum fine might provide a stronger incentive to polluters to reduce emissions, rather than increasing the number of inspections by the authorities.

In 1988, one year after the establishment of the EPA in the central government and immediately after the substantive start of a comprehensive environmental policy, it was thought that exposure of violation was the main purpose of inspections by the authorities, and violations were actually found and fines were collected in many cases (fines were collected in 35.6 per cent of inspection cases). In the second half of the 1990s when the rate of fine payment began falling, the main purpose of the inspections had most likely changed from exposure of violations to ascertaining compliance with emission standards. There has been a movement to switch inspections of factories and hog farms by the authorities to self-assessments by polluters, who report to the authorities through private inspection contractors.[7]

As seen above, the regulation of air, water and waste pollution has achieved substantial success by measures such as raising the frequency of inspections, instituting severer emission standards and raising the maximum amount of fine payments. Nevertheless, as far as the data of Taiwan as a whole is examined, the situation of air pollution has improved, but the situation of water pollution has not improved. Moreover, the situation of waste has not improved significantly.

The following can be considered as the cause. Concerning air pollution discharged by fixed sources, pollution discharged by the small, unregistered factories that authorities cannot monitor has had a significant influence, which cannot be disregarded given the large number of such factories. As a result of the efforts at control by the authorities, improvement in the pollution situation may be progressing for stationary sources of air pollution. However, it is thought that the introduction of economic instruments, pollution tax on energy use and indirect regulation was also a major factor in improvement.

Concerning mobile sources of air pollution (traffic such as automobiles and motorcycles), it is thought that improvement progressed quickly late in the 1990s due to the development of the transit infrastructure in major cities and the tightening of emission gas regulations for automobiles and motorcycles, which made the standards comparable to those of developed countries.

Concerning water pollution, the manufacturing industry and the hog farms have been the target of control. In terms of the total amount of discharge effecting BOD_5, manufacturing industry accounts for approximately 50 per cent, animal husbandry (mainly hog-raising) accounts for approximately 25 per cent and drainage from households, 25 per cent. In some river basins, drainage from households is a major source of water pollution, so even if measures for the manufacturing industry and the hog farms are effective to some extent, water quality might not be improved.

4.   Environmental policy by local government and local politics

The deployment of environmental policy was long delayed in Taiwan, given its degree of economic development. A factor that has been raised as a cause of the delay is the fact that the authority of local governments was markedly restricted under the authoritarian regime of the KMT. This contrasts with the situation in Japan during its high-growth era, where the local governments, rather than the central government, took the lead in developing measures against industrial pollution. Decentralization is not only an important element of democratization, but also the result of democratization. Unlike South Korea, which was under an authoritarian regime similar to Taiwan's, elections of the local governments' heads (except in Taiwan Province, Taipei City and Kaohsiung City) and a local assembly were held, even under the authoritarian regime of the KMT.

The KMT administration, which was a 'transplanted government', moved from mainland China after the end of the colonial rule by Japan, replaced the colonial government and did not have any political foundation in the local area of Taiwan. The KMT government succeeded in reigning over Taiwan as a whole, preventing the local political forces from gaining influence in any region by building a political network with cooperative parties across the whole of Taiwan, acquiring political influence in districts that existed from colonial days, and making two or more political factions in each region to compete with each other. Their monopoly on the central government's political power and an overwhelming monopoly of violence served to support the success of the KMT dominance.[8]

The KMT did not always have complete control of the local governments, and at least some anti-KMT forces (called the 'political outsiders') filled positions as members of local assemblies or heads of local governments even under the authoritarian system. In the process of the democratization in the 1980s, the opposition forces led by the Democratic Progressive Party (DPP) formed in 1986 and also advanced rapidly in the local governments and local assemblies. In the local election in 1997, the DPP won a majority of the head positions in county governments.

Various kinds of social movements that had been oppressed under the authoritarian regime formed relationships with each other, and the DPP, formed by the 'outsiders' in every region, kept close relations with the anti-pollution and anti-development movements, and especially with the anti-nuclear power plant movement. It has been said that their close relationships with the anti-pollution movement made the local government heads of the DPP prudent when it came to promoting large-scale development projects. For example, the Ilan county governor of the DPP was opposed to the construction project of a sixth naphtha cracking plant in Ilan by the Formosa Plastic Group, Taiwan's largest private business group, when forced to agree to the building of the plant in Ilan County. Taiwan's sixth naphtha cracking plant (called 'Liuqing') was announced in 1986, and the central government strongly supported the project.

In the discussions of studies on environmental policy in Taiwan in the early 1990s, it was expected that political liberalization and democratization would expand local autonomy and, as a result, that local government's environmental policies by would be promoted by the change in the political administration of local government from the KMT to the DPP, which was closely aligned with the anti-pollution and anti-development movements.[9] In the early 1990s, even political science experts could not imagine that a change in Taiwan's central government lay in the near future.

In this section, the counties of Taichung, Tainan and Yunlin are presented as examples; local governments gained importance through decentralization and democratization, and we will introduce the relationship between environmental policies and environmental politics in each local government. Taichung County, Tainan County, and Yunlin County are important examples in the sense that the politics of environmental administration, or the local government's 'development and environment' policy, were especially focused on one important policy subject in each region.

The anti-development faction and the governor's frustration: Taichung County

The DPP accomplished remarkable progress through the 1990s, both in the legislative elections and in the elections for the heads of a local government in each region. In particular, in the nationwide local elections in 1997, the progress of the DPP was remarkable, and the DPP candidates won head positions in the major counties and municipalities along the west coast, where economic development was already successful. A typical example of a county governor of the DPP maintaining a strong anti-development policy is seen in the incident in which a setback was suffered by a large-scale investment plant programme that involved production of titanium dioxide (TDI) by the Taiwanese subsidiary of the Beyer Company in Taichung County in 1998.

Taichung County is located on the central west coast, a prosperous commercial area where many small and medium enterprises are operating and where economic development is going comparatively well. Liao Yonglai, the governor of Taichung County at that time, was committed to the Taichung County Pollution Prevention Association, the first anti-pollution organization formed, in 1986, legitimately in Taiwan. From the time of the association's establishment, he had a close relationship with the local environmental protection movement.[10] Liao Yonglai was one of the founders of Taiwan Environmental Protection Union (TEPU), the first nationwide environmental protection association in Taiwan, and he was the general secretary of TEPU when it was established in 1988. He entered politics after that and, after serving as DPP member of the legislature, won the governorship in the nationwide local elections in November 1997. The Assembly of Taiwan Province, which was dominated by proponents of development, and the Ministry of Economic Affairs and the Committee of Economic Construction of the central government at that time, attempted to promote the large-scale investment programmes funded by foreign capital on the largest scale ever seen in Taiwan at that time. The total amount of investment in the Beyer project was approximately NT$50 billion. The Beyer Company had applied for the project to Ministry of Economic Affairs in 1994.

The local politicians who belong to the DPP organized the Anti-Beyer Movement Association to oppose this project. They pointed out to local residents the danger of highly poisonous substances leaking from a TDI production plant into the atmosphere, and they succeeded in mobilizing the residents. Also, they succeeded in gaining the cooperation of the head of the local Wuqi Township office, who belonged to the KMT and who organized active protest activities against the project. Before being elected as the governor of Taichung County, Liao Yonglai acted as a leader in the Anti-Beyer Movement.

During the Taichung gubernatorial election campaign, Liao Yonglai unsurprisingly took the prudent position with regard to this development project. He stated that whether or not the investment programme of Beyer Company is accepted should be determined by a local referendum (plebiscite). His position did not change even after winning the Taichung County governorship. The central government was not in favour of the referendum, and the representatives of the foreign enterprise association in Taiwan put pressure on the DPP's central headquarters to make governor Liao Yonglai stop the referendum.

For the DPP, this problem led to a confrontation of interests and a conflict over a values involving 'development and environment'. Before political liberalization and democratization were fully realized, there was not yet a clear differentiation between the political movement that sought political freedom and the social movement that sought to resolve social problems, including environmental problems. As a result, both movements formed a de facto coalition against the KMT regime; both sides used the political free-dom gained from the KMT through each other's efforts, and maintained inter-organizational relationships through personnel. However, at that time, it was thought that a close relationship with the anti-pollution movement might become an obstacle to the DPP in its effort to take the reins of gov-ernment from the KMT, so the DPP worked to improve its relationship with the business and the industrial world in order to expand its base of support. In response to pressure from the industrial world on the DPP headquarters, Hsu Hsinliang, the president of DPP, compromised and agreed to hold a pub-lic hearing instead of a referendum, and Chen Wenqian, a party spokes-woman, stated that an unrestricted referendum would engender unreasonable conflict. The DPP headquarters feared that some of its factions continued to take an anti-development and anti-industry position and tried to prevent Governor Liao Yonglai from taking an extreme anti-development policy. The DPP made ambiguous its position on the above-mentioned investment programme.[11]

The Beyer Company feared that the investment programme would be voted down in a referendum after all, and the DPP headquarters could not change Governor Liao Yonglai's opinion on referendum implementation. In March 1998, the Beyer Company announced that it would give up the devel-opment project, in spite of a bill, already approved by the Assembly of Taiwan Province, to lend the company a plant site in Taichung Port that was owned by the Taiwan Provincial Government. This incident shocked the industrial community of Taiwan, which had been expecting a spillover effect from the large-scale investment of the Beyer Company. Moreover, it meant that the local government, in fact, refused the development project even though the cen-tral government promoted it as a matter of national policy. In April 1998, immediately after the TDI plant was abandoned, Governor Liao Yonglai, who feared being labelled anti-development, informed the Beyer Company that investment programmes for products with lower danger levels than TDI were welcome in the county. However, the Beyer Company could not promise a new investment programme at that time.

After the stagnation of the Taiwanese economy worsened in 2000, the economy again became a political issue in the gubernatorial election in November 2001. Liao Yonglai, who was running for re-election, was adversely affected when the Taichung County Government reminded voters of the economic opportunity that had been lost by the implementation of the referendum and non-acceptance of the Beyer Company development plan. The opposition candidate from KMT also used the Beyer Company incident, in which Liao Yonglai buried the Beyer Company investment pro-gramme, with considerable effect on the election campaign because the regional economic stagnation also affected Taichung County. In the end, Liao Yonglai was not re-elected, being beaten by Huang Zhongsheng, the KMT candidate.

The lesson for environmentalists from this incident is that a powerful opposition pressure was created when the local governor gave the appearance of dogmatically refusing the investment programme by using his authority rather than forming a consensus in the community based on discussion that fully involved the residents extensively over long period of time.

The difficult political situation in which the DPP found itself at the time is also considered to have affected Liao Yonglai's failure in the Taichung County gubernatorial election. Although Chen Shui-bian was elected in the presidential election in March 2000 and the DPP succeeded in taking over the central government, following the presidential inauguration in February 2001, a political controversy arose over the fourth nuclear power plant (called He-si), where construction had been halted. The DPP, which had included opposition to the fourth nuclear power plant construction as part of its campaign platform, found itself in a deadlock in the Legislative Yuan (the Parliament), due to obstinate resistance from the opposition parties (KMT and the People First Party), which constituted a majority, although the construction halt had been determined in December 2000 after the Chen Shui-bian administration inauguration. Finally, the DPP was forced to accept the resumption of construction of the fourth nuclear power plant in February 2001.[12]

The DPP, which had become the governing party, and the environmental protection and anti-development movements, represented by anti-nuclear power plant movement, continued to have a difficult relationship for years. After his loss in the gubernatorial election, Liao Yonglai, who opposed the large-scale development project, was not re-elected to the Taichung governorship in the November 2001 election, and this reflected the political situation of those days, which saw the end of the honeymoon between the DPP government and the anti-development and environmental protection movements.

An environmental conservationist governor's dilemma: Tainan County

Chen Tangshan is an example of DPP county governor who promoted a large-scale development project in the county. Chen Tangshan, who played an active role in Taiwan's independence movement and was obliged to live in exile for a long time, went back to Taiwan after democratization. He was inaugurated as the Tainan governor after serving as a legislative member and, as governor, he promoted the Binnan Industrial Complex development project in Tainan County. This project reclaimed land from the Qigu Lagoon and tidal wetland in the Tainan County and built a large-scale industrial complex, including the iron mill of the Yieh-Loong company as the first private integrated steelworks, and the seventh naphtha cracking plant.

Moreover, Governor Chen Tangshan also strongly promoted construction of garbage incinerators in Tainan County where local residents were opposed to them due to the dioxin problem, etc., and he continued to ignore local environmental protection organizations. Although agriculture is prosperous in the Tainan County, economic development was not progressing as fast as the adjoining Kaohsiung County and Kaohsiung City.

It was Su Huanzhi, a member of the Legislative Yuan from Tainan County at that time, who was an active leader of the opposition movement against the Binnan Industrial Complex development project in Qigu, although he belonged to the same DPP as Governor Chen Tangshan. Su Huanzhi was known as a young politician on the side of environmentalists who tackled the environmental problems of southern Taiwan, in addition to being known for his involvement in the opposition movement against the Binnan Industrial Complex development project. Qigu Lagoon serves as the winter habitat of a globally scarce migratory bird, the blackfaced spoonbill, and the large majority of blackfaced spoonbills in the world wintered there.[13] Binnan Industrial Complex development project itself had fallen into difficulties. The business of the Yieh-Loong company was declining, and Tuntex group, which was going to be the operator of the seventh naphtha cracking plant, lapsed into financial difficulties due to the recession of the Taiwanese economy, which became dire from around 2000.

In the county gubernatorial election in November 2001, Chen Tangshan completed his term of office and Su Huanzhi stood instead as the DPP's gubernatorial candidate. The candidate Su Huanzhi made a campaign pledge to stop the Binnan Industrial Complex development project to which he had been opposed for some time. Until the last phase of the electoral campaign, Governor Chen Tangshan opposed Su Huanzhi's candidacy and did not show a positive attitude towards helping his campaign, even though they belonged to the same DPP and Su Huanzhi was to be his successor. In the end, Su Huanzhi won the Tainan governorship. Moreover, Chen Tangshan, who had finished his term as governor, stood as a candidate and won a seat in the national legislature as the representative from Tainan County. In this way, the positions of Chen Tangshan and Su Huanzhi were completely reversed.

As governor, Su Huanzhi used his political power and ended the Binnan Industrial Complex development project, which was already suffering a setback due to the deterioration of economic conditions. Chen Tangshan, who returned to the national political arena as a legislative member, continued to severely criticize the anti-development policy of Governor Su Huanzhi.[14] However, Su Huanzhi, who was initially regarded as an environmental conservationist, proceeded to promote the construction of the Tainan International Airport, as governor of Tainan County. In this plan, a part of Qigu Lagoon was to be developed, and many environmental protection organizations and local residents had expressed their opposition. Moreover, the extension of the Kaohsiung International Airport and the expansion of the international airlines there were already advanced, so construction of another international airport in the adjoining Tainan County was of questionable necessity in southern Taiwan.

A complicated political situation serves as a backdrop for the promotion of this large-scale development project, i.e. the Tainan International Airport development project, by Su Huanzhi, who heretofore was known as a champion of the anti-development faction. It is difficult to understand the situation simply in terms of confrontation between development promoters and environmental conservationist, and 'defection' of an environmental conservationist.

Su Huanzhi's contradictory and complex attitude towards 'development and environment' issues could be explained as follows. He gave priority to his position as governor rather than to his position as an environmental conservationist; he was opposed to Binnan Industrial Complex development project not only because he was an environmentalist, but also because he could not allow a major private company to have a tight relationship with KMT and thereby increase its power around Qigu, where his election base was located.

The following factors may also explain his position on development in Tainan. Governor Su Huanzhi did not want to oppose, as much as he could, development projects that the central government promoted in Tainan, which was also the home county of President Chen Shui-bian. Also, while local economic conditions were worsening, he did not want to cause a backlash from the business community.

Local government as promoter of development projects: Yunlin County

Yunlin County is presented as an example of a county where the local government is more eager for development projects than for environmental protection. Yunlin County is an underdeveloped area; the level of industrialization is exceptionally low on the west coast, and the major industries were agriculture, hog farms and construction, until recently. The political base of the KMT in Yunlin County is strong, and the DPP was unable to win the gubernatorial election there until 2005. In the gubernatorial election following the death of Governor Su Wenxiong in November 1999, the KMT was divided, and Chang Rongwei, who could not obtain official recognition of the KMT but stood as an independent candidate, won in the end. Subsequently, in order to obtain more votes in the presidential election in March 2000, the KMT accepted the return of Governor Chang Rongwei to the party. Chang Rongwei, who won the governorship and also attracted support for the KMT, drove other factions from Yunlin for political domination of Yunlin County, expanded his influence and was re-elected as governor in November 2001.

In Yunlin County, which is relatively behind in development, economic development has been furthered in various ways in recent years, and the problems of destruction of scarce natural resources and industrial pollution have surfaced.[15] It is said that Chang Rongwei has a close relationship with construction companies. He ignored the local residents' movement, which requested protection of a globally scarce migratory bird, the fairly pitta, which inhabits Zhentoushan Mountain in Huben Village, Linnei Township, and he did not attempt to take any measure at all against the contractor who tried to shave off the mountain in order to extract sand for construction work.[16]

Moreover, the Legislative Yuan and the EPA opted for an expenditure freeze on the construction budget for the garbage incinerator construction project, which had been furthered by BOT in Linnei Township of Yunlin County, as an exception to the policy to build at least one incinerator in each county, which had been promoted by EPA until then. However, opposing this freeze by the central government, the Yunlin County Government forced ahead the construction of the incinerators by using the BOT procedure and financing instead of the central government's budget. The construction of a water purification plant to supply water to surrounding counties and cities is planned near the incinerator, and there is fear over contamination of the water source of the water purification plant.

Besides pollution problems mentioned above, the economic efficiency of the incinerators was considered to be questionable. Therefore, many local residents, some of the village heads in Linnei township and environmental protection organizations continued to strongly oppose construction of incinerators. In addition, there has been surplus capacity in the incinerators already built in the adjoining counties and municipalities in recent years. In short, there was already an oversupply of incinerators in Taiwan.

Due to this incinerator construction problem, opposition factions and environmentalists developed a movement to recall the head of Linnei Township, who was a promoter of incinerators. The first recall vote ever held in Taiwan for the dismissal of township head by direct vote of the residents was held in September 2003, but the recall was not successful in the end.

Moreover, in the Meilin district in Douliu City, an opposition movement developed against a dam construction project. In this project, the construction plan was advanced through a poor-quality environmental impact assessment that led to the approval of the project and disregarded the existence of an active fault under the planned site.

Furthermore, in Huwei Township a severe soil pollution problem was revealed in 2001 that involved cadmium, which had been discharged over many years from the drain of a chemical factory that adjoined farmland. At the time the problem was discovered, the factory had already converted its production process and was no longer discharging cadmium. Since no statute regulating the discharge of cadmium had existed when the discharge was made, no administrative penalty or criminal liability were imposed on the polluter company. Farmers who suffered damage from cadmium pollution could not receive proper compensation for their contaminated farms.

If the local government of Yunlin County, whose authority was strengthened and whose source of revenue and staff were expanded by decentralization, had been doing its best, such serious environmental problems should not have occurred. However, the Yunlin County Government never accepted the local community's demands for environmental protection and instead stuck consistently to its position of promoting development. Moreover, the Yunlin County Government did not attempt to solve the existing environmental pollution problems. In the waste incinerator construction problem in Linnei, in spite of the central government's having admitted the problem and having stopped the plan, the local government resisted by using its own authority and the BOT procedure, thereby reversing the stoppage of the incinerator construction.

After the setback in Ilan County, the sixth naphtha cracking plant of the Formosa Plastic Group was finally built in Mailiao Township of Yunlin County. In Yunlin County, which was behind in development, large-scale development has just started in recent years. There may be no way to avoid promotion of economic development by local governments in areas lagging in development. However, even if development yields short-term profit, neither the development that does not consider environmental protection nor the development that destroys scarce natural resources will make profit for the community in the long run.

The reason the Yunlin County Government favours promotion of development has been thought to be due to the governor's deep connection with construction companies' interests and strong connection with vested local interests. The central government cannot take effective countermeasures in such a situation. The Chen Shui-bian administration approached Governor Chang Rongwei and tried to win him over to President Chen's side in order to obtain more votes through the cooperation of the governor in Yunlin County in the presidential election on 20 March 2004, by providing new public construction projects.[17] Given that the central government was approaching the county governor who promoted development in league with vested interests, it could hardly be expected that the central government would champion environmental protection in Yunlin County and apply brakes to the governor's policy of placing priority on development.

As stated previously, the KMT prevented continuous rule over a long period in any district by a specific political faction. The KMT's authoritarian regime, which continued until the middle of the 1980s, achieved this by making two or more factions compete in each district. However, the KMT was unable to control local governments in such a way after democratization. Political liberalization and democratization progressed under the Li Tenghui administration of the KMT from 1988–2000.

On the other hand, the underground influence, which then began to appear overtly on the political stage, openly pressured politicians and strengthened its political influence; furthermore, members of the underground began to appear themselves as politicians. Although KMT still imposed an authoritarian regime, the political liberalization under the Li Tenghui administration allowed the underground, which had been suppressed by the state, to begin to intervene in politics openly. The problem of underground influence was considered to be the worst by-product that democratization of Taiwan brought about.

In Yunlin County also, the KMT's authoritarian regime made two or more local political factions compete and forced false changes of power, thereby cleverly avoiding the situation where a specific local faction governed the whole county over a prolonged period of time. This power of KMT to control local politics faltered after democratization progressed in Yunlin County. Although Chang Rongwei left the KMT after the by-election of a governor in 1998, he returned to the KMT after success in a gubernatorial election. Chang Rongwei strengthened his political base by approaching the Chen Shui-bian administration. The leading counterbalancing force in the county was lost until early 2004.[18]

As seen above, democratization brings about decentralization, but the local governments' environmental policies in Taiwan could not be comprehended only by using a simple diagram where decentralization was expected to promote development of local environmental policies. The conditions of local politics and local governments in each area, and their environmental and natural resource conservation policies, should be closely related to the stage of economic development. Issues of 'development and the environment' can only advance through adjusting the various interests of parties involved in the political process. The local government's environmental policy may also be affected by the underground influence, which is a by-product of democratization, as shown by the example of Yunlin County.

Moreover, the example of the Taichung County also suggests the possibility that the local government's efforts to promote discussion and consensus-building in the community were not sufficient.

5 Conclusion

In this chapter, the institutional progress in the environmental administration of Taiwan was clarified first, then the state of environmental quality improvement was evaluated and the effectiveness of the pollution emission regulations by the environmental policy administration was evaluated. Moreover, the various influences of decentralization that accompanied democratization on the local governments' environmental policies are analysed.

We observed progress in the organizations and institutions involved in environmental administration and considerable success in implementing regulations on pollution emissions at their source. Although improvement was generally found in air pollution, no tendency towards improvement of the water pollution problem was witnessed, to the extent that the pollution situation was observed in major rivers.

The political situation of environmental policy is complicated. This chapter focused on political situations where there were conflicts of interest in policies connected with 'development and environment' that exerted particular influence on local governments, as seen the cases in the counties of Taichung, Tainan and Yunlin.

Although the KMT administration of Taiwan, which had justified oppression of political freedom by promoting economic development, succeeded in industrialization as a result, it brought about delay in countermeasures for environmental pollution problems generated due to industrialization, caused expansion of pollution and implementation of unsuitable large-scale development projects.

Expression of citizens' dissatisfaction with environmental pollution was gradually realized for the first time in the process of political liberalization and democratization, and the social pressure for environmental conservation advanced the development of environmental administration of the government and pollution prevention measures of companies.

The government had neither the organization nor the experience required to solve environmental problems and was not able to work effectively in the disputes which occurred frequently in various places when the dissatisfaction of citizens with environmental pollution first began to explode.

Since citizens' complaints about environmental degradation and protests over pollution were suppressed by violence before political liberalization progressed, neither such systems nor organizations were needed at all.

Such an 'institutional vacuum' brought about the social confusion and made resolution of pollution problems by 'self-relief' movements popular around the country. The 'institutional vacuum' continued from the second half of the 1980s to the early 1990s.[19] However, a significant improvement in environmental pollution was seen in some fields as people's dissatisfaction exerted pressure for the development of a system of environmental organizations and institutions. More and more frequently, solutions to environmental problems were sought through environmental administration organizations, such as those in local government, instead of through self-relief (Figure 9.3).

However, when we started analysis of environmental policy in Taiwan in the early 1990s, we could fully understand neither the movement towards an environmental administration after the change of power in the central government, nor the diversification of the local governments' environmental policies, after the middle of 1990s by using only the simple diagram where political liberalization and democratization promoted the progress of environment policy.

After examining the context of each situation in detail, the factors that promoted or impeded environmental policy need to be comprehensively analysed. Decentralization made it impossible for the KMT to control the local governments of every region fully from the centre. The interests peculiar to each area made conspicuous the political confrontations in connection with each policy subject and generated a vastly different political situation for every area.

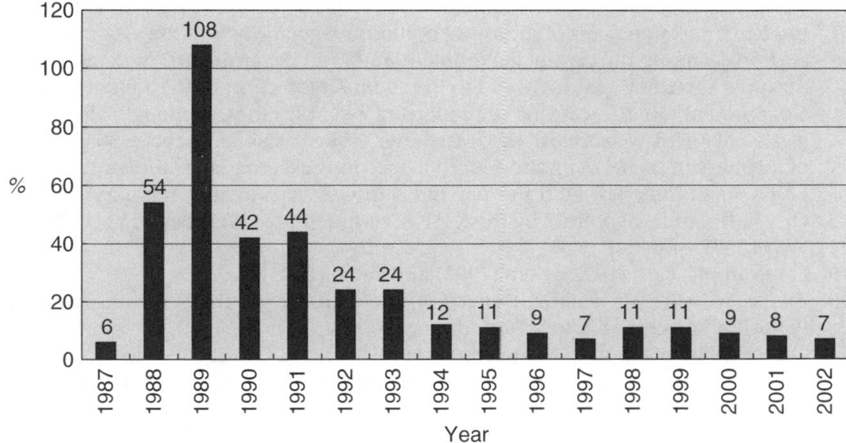

Figure 9.3   Number of significant environmental disputes
Sources: Environmental Protection Administration, 'Pollution Dispute Resolution White Paper';
Environmental Protection Administration, Executive Yuan (2000).
Note: 1987 figure is from July to December only.

The conflicts of interest peculiar to each area involving 'development and environment' came to have a large influence also on the local governments' environmental policy. Furthermore, changing factors in the conflicts of interest involved in 'development and environment' shook the foundation of local politics in some areas severely. Thus, policies involving 'development and environment' in Taiwan are entering a new stage.

Notes

1. On environmental pollution and environmental policy in Taiwan up to the early 1990s, see Terao (1993) for detail. The load on environmental resources in Taiwan at that time was quite large, compared to other developing countries, and even compared to developed counties except Japan (see Table 6.1 of Terao, 1993). See also, Sato (1992).
2. Fan (2000) described networks among social movements including environmental protection movements in the process of democratization. Also, Ho (2000) analyses political aspects of environmental protection movements of Taiwan in detail. See also, Chang (1993) and Terao (2002a).
3. Ho (2000), Tang and Tang (1997) and Terao (2002b) analyse causes of 'self-relief' pollution disputes.
4. The process of 'citizen lawsuits' should be closely related to that of administrative litigation. However, the process of administrative litigation is not yet organized and recently was reviewed for reform. Based on an interview with Professor Liu Zhongde (Department of Law, National Chengchi University) on January 2004.
5. According to those data, the quality of air and water has not changed significantly since evaluation by Terao (1993) in the early 1990s.

6. Environmental Protection Administration (1999), for example, explained the implementation process of industrial pollution regulations and the assignment of authority among the central government and local governments, in detail.
7. According to the Department of Environmental Protection of the Tainan County Government, in the case of self-reporting by hog farms, although almost all levels measured were below the wastewater effluent standard, about 30 per cent of the measurements of unannounced inspections exceeded the effluent standard. Based on an interview with the Tainan County Government in February 2003.
8. On control of local politics by the KMT government, see Wakabayashi (1992) and Wu (1987) for detail.
9. For example, Sato (1992), Ueta (1993) and Terao (1993).
10. Interview with Mr Huang Dengtang, the founder of the Taichung County Pollution Prevention Association, in May 2000.
11. On the DPP's reaction to the referendum in Taichung County, see Ho (2001).
12. See Ho (2003) on the political process within the DPP concerning issues in the construction of the fourth nuclear power plant.
13. On conservation movements to save the blackfaced spoonbill in Tainan, see Chen and Ueta (2000) for detail.
14. Chen Tangshan became the Minister of Foreign Affairs on May 2004 due to the Cabinet reshuffle of Chen Shui-bian administration.
15. Most of environmental pollution and resource degradation issues in Yunlin County are based on field research in Yunlin in November 2002 and February 2003.
16. Tang and Tang (2003) studied the conservation movement to save the fairy pitta in the village of Huben.
17. See Ogasawara (2000, 2003 and 2004) on local politics in Yunlin County and Governor Chang Rongwei's dominance there.
18. Governor Chang Rongwei was placed on the wanted list on 28 August and arrested on 10 December for corruption related to the incinerator construction project in Linnei Township.
19. See Terao (2002b) concerning the 'institutional vacuum' in the democratization process in Taiwan.

References

Chang, Mau-kuei (1993) 'Middle class and social and political movements in Taiwan: Questions and some preliminary observations,' in Mau-kuei Chang (ed.) *Discovery of Middle Class in East Asia*, Taipei, Institute of Sociology, Academia Sinica, pp. 121–76.

Chen, Li-chun, and Kazuhiro Ueta (2000) 'Taiwan,' in Nihon Kankyo Kaigi (Japan Environmental Council) (eds), *Ajia Kankyo Hakusho (The State of Environment in Asia) 2000/2001*, Tokyo, Toyo Keizai Shinposha, pp. 253–62 (in Japanese).

Environmental Protection Administration (1999) *Study on Reform of Environmental Regulation Implementation (The End of Term Report)*, Environmental Protection Administration, Executive Yuan (in Chinese).

Fan, Yun (2000) *Activists in a Changing Political Environment: A Microfoundational Study of Social Movements in Taiwan's Democratic Transition, 1980s–1990s*, (Ph.D. dissertation), New Haven, CT., Department of Sociology, Yale University.

Ho, Ming-sho (2000) *The State and Democratic Society in Democratic Transition: Case of Environmental Movement in Taiwan, 1986–1998*, (Ph.D. dissertation), Taipei, Department of Sociology, National Taiwan University (in Chinese).

Ho, Ming-sho (2001) 'Institutionalization of Social Movements: Case of Environmental Movement in Taiwan, 1993–1999,' paper presented at the Conference on 'Organization, Identity, and Activist: Study on Social Movement in Taiwan,' Taipei, Institute of Sociology, Academia Sinica, 21 June (in Chinese).

Ho, Ming-sho (2003) 'Politics of anti-nuclear protest in Taiwan: A case of party-dependent movement (1980–2000),' *Modern Asian Studies*, 37(3) pp. 683–708.

Ogasawara, Yoshiyuki (2000) 'Unrinken kara mita 2000nen soto senkyo' (Presidential Election in the Year of 2000 from the view of Yunlin County), (http://www.tufs.ac.jp/ts/personal/ogasawara/paper/yunlin2000.html) (in Japanese).

Ogasawara, Yoshiyuki (2003) '2004nen soto senkyo no mitooshi (I)', (A Perspective of Presidential Election in the Year of 2004), (http://www.tufs.ac.jp/ts/personal/ogasawara /prospects2004.html) (in Japanese).

Ogasawara, Yoshiyuki (2004) 'Taiwan chiho seiji no doko: Unrin ken Chang Rong-wei kencho no maturo,' (Trend in Taiwan Local Politics: Last Days of the Governor, Chang Rong-wei in Yunlin County), (http://www.tufs.ac.jp/ts/personal/ogasawara/paper/zhangrongwei2004.html) (in Japanese).

Sato, Yukihito (1992) 'Taiwan: Kaihatsudokusai no fu no isan.' (Taiwan, the Negative Legacy of Developmental Dictatorship) in Shigeaki Fujisaki (ed.), *Hattentojokoku no Kankyo Mondai (Environmental Problems in Developing Countries)*, Tokyo, Institute of Developing Economies, pp. 62–77 (in Japanese).

Tang, Ching-ping and Shui-yan Tang (1997) 'Democratization and Environmental Politics in Taiwan,' *Asian Survey*, 37(3) pp. 281–94.

Tang, Shui-yan and Ching-ping Tang (2003) 'Local governance and environmental conservation: gravel politics and the preservation of an endangered bird species in Taiwan,' *Environment and Planning A*, 35.

Terao, Tadayoshi (1993) 'Taiwan: Sangyo kogai no seijikeizaigaku,' (Taiwan: Political Economy of Industrial Pollution) in Reeitsu Kojima and Shigeaki Fujisaki (eds), *Kaihatsu to Kankyo: Higashi Ajia no Keiken (Development and the Environment: Experiences of East Asia)*, Tokyo, Institute of Developing Economies, pp. 139–99. (in Japanese).

Terao, Tadayoshi (2002a) 'Taiwan: From Subjects of Oppression to the Instruments of "Taiwanazation",' in Shin'ichi Shigetomi (ed.), *The State and NGOs: Perspective from Asia*, Singapore, Institute of Southeast Asian Studies, pp. 263–87.

Terao, Tadayoshi (2002b) 'An Institutional Analysis of Environmental Pollution Disputes in Taiwan: Cases of "Self-Relief",' *Developing Economies*, XL, (3) pp. 284–304.

Ueta, Kazuhiro (1993) 'Taiwan no kankyo seisaku to nihon moderu,' (Taiwan's Environmental Policy and the Japanese model) in Reeitsu Kojima and Shigeaki Fujisaki (eds), *Kaihatsu to Kankyo: Higashi Ajia no Keiken (Development and the Environment: Experiences of East Asia)*, Tokyo, Institute of Developing Economies, pp. 253–67 (in Japanese).

Wakabayashi, Masahiro (1992) *Taiwan: Buretsukokka to Minshuka (Taiwan, Decentralization and Democratization)*, Tokyo, University of Tokyo Press (in Japanese).

Wu, Nai-teh (1987) *The Politics of Regime Patronage System: Mobilization and Control within an Authoritarian Regime*, (Ph.D. dissertation), Chicago, IL., Department of Political Science, The University of Chicago.

10
Environmental Cooperation in East Asia: Comparison with the European Region and the Effectiveness of Environmental Aid

Hideaki Shiroyama

Introduction

One of the important characteristics of any effective environmental policy is local information and local response. We cannot enforce an international environmental policy supranationally from above; we have to build an international environmental regime based on spontaneous local responses. So the regional dimension of the environmental policy is very important for international environmental governance. In practice, regional environmental regimes have been emerging in Europe, North America and East Asia since the 1970s. The regions have also been fields for experiments in innovative methods of international environmental governance. Japan is one of the major actors promoting an environmental cooperation regime in East Asia.

In this chapter, I would like to analyse the characteristics of the environmental cooperation regime of the East Asian region in comparison with the environmental cooperation regime of Europe. Based on that comparison, recent initiatives for environmental cooperation in East Asia since the mid-1990s will be analysed.

In addition, I would also like to analyse the effectiveness of regional environmental cooperation in East Asia, focusing on environmental aid as a major policy instrument. Japan, directly or indirectly (through the World Bank and Asian Development Bank [ADB]), tries to promote the development of environmental cooperation by the use of environmental aid. Bilateral and multilateral aid programmes are separate from the formal process of setting up the regional environmental cooperation regime; however, these aid programmes are closely related to setting up the regime, and interaction between various actors involved in such aid highlights the important aspects of the operation of the regional environmental regime.

As a case study of environmental aid in East Asia, this chapter will focus on Japanese aid to China. Since China, which is a major source of emissions in East Asia, is one of the developing countries having a great impact not only on the regional economy but also on the global economy, international aid is necessary, at least transitionally. With the great concern about China's environmental issues felt by the Japanese people and government, Japan has been a major donor of environmental aid to China.

Before entering the concrete analysis, I would like to make one point. In the following analysis, I use concepts such as ad hoc regime and comprehensive regime. An ad hoc regime is a sector-specific or subregional regime. On the other hand, a comprehensive regime is a regime managed by an economy-oriented but multisectoral (or general purpose) organization such as the EC (European Community)/EU (European Union) or APEC (Asia Pacific Economic Cooperation). The historical and the operational relationship (inter-organizational relation) between the ad hoc regime and the comprehensive regime is one of the major focuses of the comparative analysis.

1. Comparison of regional environmental regimes: characteristics of environmental cooperation in East Asia

The historical development of the environmental cooperation regime in the European region: the bottom-up approach

The development of the ad hoc regimes

In Europe, the development of the ad hoc regimes preceded the effective development of the comprehensive regime.[1] The regional sea regimes and the acid rain regime can be cited as examples of ad hoc regimes. In the process of building these regimes, Scandinavian countries were strong supporters. These ad hoc regimes were not only established effectively prior to the comprehensive environmental regime, but also played important roles in the actual development of the comprehensive environmental regime. In that sense, it is possible to say that the environmental regime in the European region has been built based on the bottom-up approach.

Regional sea regimes   In Europe, each regional sea developed a regional sea regime on an ad hoc basis. For example, the North Sea and the Baltic Sea developed regional sea environmental regimes based on the initiatives of Scandinavian countries (Haas, 1993).

The Convention for the Prevention of Marine Pollution by Dumping from Ships and Aircraft (Oslo Convention) to protect the Northeast Atlantic was signed in 1972 by 13 governments. Twelve Northeast Atlantic governments also signed the Convention for the Prevention of Marine Pollution from Land-Based Sources (Paris Convention) in 1974. The Oslo Commission (OSCOM) was established in 1974 to control the marine dumping in the Northeast Atlantic, and the Paris Commission (PARCOM) was established in 1978 to control the land-based sources of Northeast Atlantic pollution. OSCOM and PARCOM shared a joint secretariat in London (OSPARCOM). In addition to that, seven Baltic governments signed the Convention on the Protection of the Marine Environment of the Baltic Sea Area (Helsinki Convention) in 1974. The Helsinki Commission (HELCOM) was established in 1980 to control all sources of Baltic marine pollution.

For the next ten years, they developed their own methods for environmental protection through practical activities in regional sea regimes. Participants in the managing commissions, that is, OSCOM, PARCOM and HELCOM, were middle-level experts or practitioners, so their activities were low profile.

However, the political environment surrounding these regional sea regimes changed dramatically in 1980s. Behind this change were environmental shocks, that is, the widely publicized 1982 forest death in West Germany and the 1986 Chernobyl disaster in the USSR. Environmental issues became the politically salient issues. In this political environment, in 1984, the North Sea Ministerial Conference was established. Participants in this conference were at the ministerial level, and their activities were high profile. This North Sea Ministerial Conference was also convened in 1987 and 1990. These ministerial conferences produced Ministerial Declarations, which were not legally binding. However, in practice, OSCOM and PARCOM produced binding regulations in conformity with those Ministerial Declarations, so the ministerial conferences had de facto binding power, backed by public opinion.

Through these processes, several policy innovations were introduced. First, the 'precautionary principle' was established in the 1989 PARCOM recommendation. This concept originated in the German domestic environmental policy. It reflected a change of policy focus from ex post reparation to preventive regulation, and a shift of the burden of proof from the injured to the assaulter. This concept was first articulated internationally at the North Sea Ministerial Conference, and through these ministerial conferences, it was exported to the EC where the 'precautionary principle' was put into the Maastricht Treaty. The concept of BAT (best available technology) was also developed through a similar course.

Second, the method of across-the-board cuts was introduced in 1987. An across-the-board cut of 50 per cent was instituted for 37 significant pollutants, with a 70 per cent cut for dioxins, mercury and cadmium. Until then, the cuts had been applied on a substance-by-substance basis, which led to a disorganized and incoherent set of policy efforts, and there had even been disagreement about the emission standards because the UK had been advocating quality standards.

Third, strict decisions were reached on marine dumping, which stated that 'no material should be dumped after January 1, 1989, unless there are no practical alternatives on land and it can be shown to the competent international organizations that the materials pose no risk to the marine environment'.

Acid rain regime   The acid rain regime in Europe was also established on an ad hoc basis (Levy, 1993). Norway and Sweden became concerned with the acid rain issue because their fish stocks were threatened. They focused on that issue at the 1972 UN Conference on the Human Environment. After that conference, the OECD (Organization for Economic Cooperation and Development) launched a study to measure the long-range transport of air pollution. OECD established a network of the monitoring stations, which later in 1978 was given institutional status as the Cooperative Programme for Monitoring and Evaluation of the Long-range Transmission of Air Pollutants in Europe (EMEP).

On the other hand, this acid rain issue was discovered politically through another path, that is, through the CSCE (Conference on Security and Cooperation in Europe) and the ECE (UN Economic Commission for Europe). The CSCE was a security regime which was established in 1975 to institutionalize the 'detente' between the Eastern countries and the Western countries, mainly in Europe. The CSCE was trying to build confidence among countries using a non-security agenda. Based on the Soviet's proposal, environmental, energy and transport issues were chosen as candidate topics at the high-level East–West meeting. The task of choosing the specific topic fell to the ECE. It contained virtually the same membership as the CSCE but, unlike the CSCE, the ECE had the organizational infrastructure. The ECE studied the options during 1977–8 and chose the environment as the best candidate for the East–West conference. Finally, air pollution was specified as a suitable topic for the international convention. After the negotiation during 1978–9, the LRTAP (Long-range Transboundary Air Pollution Convention) was signed in November 1979.

The Air Pollution Unit of the ECE was assigned the secretariat's function, and an Executive Board was established for the LRTAP. Initially, their activities were mainly scientific research and were seen as a continuation of OECD (EMEP)'s activities; but from the mid-1980s, they began to negotiate regulatory protocols. These regulatory protocols were binding to the parties but the method of implementation was entrusted to each party's discretion. In 1985, the Sulphur Protocol was signed, which called for 30 per cent across-the-board reductions. It had been sought by the Scandinavians since 1978, but the change of German attitude in early 1980s was decisive in the process towards the signing of the Sulphur Protocol. Furthermore in 1985, the Scandinavians began seeking a nitrogen oxides control, and in 1988 the Nitrogen Oxides Protocol was signed. It called on countries to freeze their nitrogen oxides emission at the 1987 level by 1995. However, 12 countries that were parties to the protocol went further by pledging to reduce nitrogen oxides emission by 30 per cent by 1998.

Role of the comprehensive regime: the EC/EU On the other hand, the comprehensive regime – the EC – began to work in environmental policy relatively recently (Johnson and Corcelle, 1989).

While the word 'environment' did not appear in the 1957 Treaty of Rome which established the EC, during the mid-1960s the EC began to recognize that the establishment of a common market also requires the enactment of the common environmental regulations. In 1967, the EC adopted its first environmental directive (concerning standards for classifying, packaging and labelling dangerous substances), and in 1970 the EC also adopted a directive on automotive emissions. In October 1972, at the Paris Summit, the EC heads of state called on the European Commission to draw up a Community environmental policy and authorized the European Commission to establish a separate administrative body dealing with environmental protection. Following that communiqué, in 1973, the EC's first Environmental Action Programme was adopted by the Council of Ministers. The second and third Environmental Action Programmes were also adopted in 1977 and in 1983, respectively. Between the early 1970s and the mid-1980s, the EC issued 120 regulations and directives, the focus of which covered wide areas, including air, water, noise, waste disposal, prevention of accidents, safety requirements for chemicals, environmental impact assessments and so on. However, compared to ad hoc regimes, the level of commitment of the EC towards environmental policy in this period was rather weak and passive.

In the mid-1980s, the situation of environmental policy changed dramatically also in the EC. The European Council at head-of-state level began to take up environmental issues more frequently. In 1983, in response to the death of the forest in West Germany, the European Council called for urgent action concerning the forest. In 1985, the European Council declared that the environmental protection policy was the fundamental component of the economic, industrial, agricultural and social policies.

In the SEA (Single European Act), which was signed in 1986 and aimed to establish a single market, Article 100A explicitly recognized the improvement of the quality of the environmental as a legitimate EC objective in its own right. Article 130R further declared that 'environmental protection requirements shall be a component of the Community's other policies'. These words reflected the change in the attitude of the EC towards environmental policy. These changes were stimulated in part by the USSR's Chernobyl disaster, the massive chemical spill of toxin into the River Rhine in 1986 and the response of the public to those mishaps.

Two points should be made clear about the environmental activities of the EC. The first point concerns the relations between the EC and the ad hoc regimes. In the case of the North Sea regime, the EC was a party to OSCOM and also participated in the activities of PARCOM. However, the attitude of the EC in the 1970s was negative towards environmental policy because the EC's interest was oriented towards economic growth. The EC exercised a major restraint on the agenda-setting and forced non-decisions in the North Sea regime. The EC Commission asserted its competence to speak for the EC where prior EC directives existed or where similar EC discussion was under way. For example, North Sea states postponed their discussion on PCBs for five years and abandoned talks on the environmental impact assessment when they found it impossible to reach an agreement on a text more stringent than the existing EC Directives. In those situations, the leading countries of Scandinavia invented an inter-organizational strategy to cope with the conservative attitude of the EC. They tried to get the agreement first from the North Sea Ministerial Conference, OSCOM, PARCOM and HELCOM countries, and after that, they brought the agenda to the EC. As I mentioned before, due to this strategy, the concepts of 'precautionary principle' and 'BAT' were adopted by the EC. However, since the mid-1980s, the EC itself has become active in environmental policy because of the politicization of the environmental issues. For example, in the third Environmental Action Programme adopted in 1983, the EC actively committed itself to the environmental protection of the North Sea. Inter-organizational competition between the EC and the ad hoc regimes was the catalyst for introducing the policy innovations.

The second point concerns the role of the legal mechanism in the EC environmental regime. The EC issued many regulations and directives, and it can be said that the EC put emphasis on legal instruments. In resolving the conflict over the interpretations of these legal instruments, the European Court of Justice (ECJ) played an important role. One of the topics, which is much discussed recently, is the relation between trade regulation and environmental regulation. Article 130T of the SEA explicitly granted member states the right to maintain or introduce national environmental standards that were stricter than those approved by the EC, provided that they did not constitute a form of 'hidden protectionism' and were compatible with the Treaty of Rome. So it is the focus of many disputes whether national environmental regulations that are stricter than EC regulations are not a form of 'hidden protectionism' or are compatible with the trade regulations of the Treaty of Rome. For example, in the 1988 decision of the European Court in the European Commission versus Denmark case (Case 302/86), the Court upheld the legality of the most important provisions of a Danish bottle recycling law (stating that containers for beer, soda, lemonade and gaseous mineral waters have to be collectable and reusable through a system of deposit and return), even though this legislation made it more difficult for non-Danish bottles to gain access to the Danish market.

The historical basis of the environmental cooperation regime in the East Asian region: the top-down approach

The inadequacy of ad hoc regimes

In the East Asian region, the development of the ad hoc regimes has been inadequate compared to the European region. This has been especially true for the Northeast Asian region (Japan, China, Korea, etc.), (Shiroyama, 2001).

Historically, there were some active ad hoc regimes even in the East Asian region. For example, the regional sea regimes in ASEAN (Association of South East Asian Nations) and in the South Pacific have been rather active. In addition to the regional sea programme, ASEAN has been engaged in many environmental programmes. There are also some attempts to establish sectoral regimes through ESCAP (Economic and Social Commission for Asia Pacific), one of the UN regional commissions.

However, as the following cases show, these ad hoc regimes in East Asia were peripheral phenomena; they were not active compared to their counterpart regimes in the European region and did not played important roles in the actual development of the comprehensive environmental regime in East Asia.

Regional sea regimes   The IMO (International Maritime Organization) and UNEP (United Nations Environment Programme) have for many years actively cooperated in encouraging the development of the regional sea regimes for combating the marine pollution. In the late 1980s, there were 11 such regional sea programmes. The East Asian Seas regime in Southeast Asia and the South West Pacific regime are two of those.

The regional action plan in Southeast Asia was first drafted in 1979 and was eventually adopted by the five ASEAN states in 1981 at the intergovernmental meeting convened by UNEP in Manila. The action plan consists of environmental assessment and environmental management components, the latter concentrating on the control of pollution from oil and from land-based sources. Subsequently, a trust fund was established and UNEP was entrusted with its management. The East Asian Seas Action Plan did not include a legal component. However, because the parties to that action plan were members of ASEAN, ASEAN provided a general framework for cooperation in marine pollution incidents in the form of the ASEAN Contingency Plan, which became effective in 1976. The focus of the East Asian Seas Action Plan was two principal tanker routes, that is, the VLCC route through the Malacca and Singapore straits, and the VLCC route through the Lombok/Makassar straits and the Celebes Sea.

In 1982, the Conference on the Human Environment in the South Pacific held in Rarotonga adopted the Action Plan for the Region, which includes environmental assessment, environmental management, legal components and institutional and financial arrangements. Although it was recognized that the South Pacific region was not traversed by the major tanker routes, the great majority of the countries of the region expressed concerns about the effects on their resources of a major spill from vessels delivering oil supplies in the region, and there was great interest in contingency planning. The impact of marine oil spills on the sensitive coastal environments of the region was identified as one of the initial areas requiring environmental assessment. In 1986, the Plenipotentiary Conference adopted the Convention for the Protection and the Development of the Natural Resources and the Environment of the South Pacific and also adopted the related protocol on the prevention of pollution by dumping, as well as on cooperation in combating pollution emergencies. The Protocol provided that the SPC (South Pacific Commission) would carry out, through the secretariat of the SPREP (South Pacific Region

Environmental Programme), a number of specific functions relating to the implementation of the Protocol. These functions included technical assistance, training, dissemination of information and checking/maintaining the emergency response communication system, etc.

ASEAN   ASEAN has been rather active with regard to environmental policies (ASEAN Secretariat, 1995). With the assistance of UNEP, ASEAN elaborated the First ASEP (ASEAN Environment Programme: 1978–82). Later, ASEP2 (1982–7) and ASEP3 (1988–92) were also planned and implemented. The focuses of these programmes were nature conservation, industry and the environment, the urban environment, the marine environment, environmental education, training and information.

Organizationally, an expert group on the environment was established under the auspices of the ASEAN science and technology committee. This expert group was renamed ASOEN (ASEAN Senior Officials on Environment). In April 1981, the first Meeting of ASEAN Environment Ministers was held. This ministerial meeting was institutionalized and has been held every two or three years since then. Compared to the North Sea region, where the North Sea Ministerial Conference was established in 1984, the establishment of the Meeting of ASEAN Environment Ministers in ASEAN occurred surprisingly early.

In April 1994, the Sixth ASEAN Ministerial Meeting on Environment was held to coordinate ASEAN's views and responses on the post-UNCED (United Nations Conference on Environment and Development) issues. This resulted in the adoption of the ASEAN Plan of Action on the Environment, which was a successor of the ASEP and which would provide ASEAN with a blueprint for its strategic policy framework, as well as a set of activities to be implemented and accomplished in the environmental sector over next four years. The harmonization of environmental quality standards (i.e., long-term goals among member countries regarding ambient air quality and river water quality and regarding the minimization of transboundary movement of hazardous wastes within the region) was also discussed in it. Those standards were general quality standards, in contrast to the specific, quantitative, across-the-board reduction standards that were used in the European ad hoc regimes.

ESCAP and ECO-ASIA   There were also some exploratory attempts to establish sectoral regimes.

In 1978, ESCAP established the Environment Coordination Unit with the assistance of UNEP. The unit was enlarged in 1989. The focus of ESCAP activities was information-gathering and agenda-setting. For example, in 1988, in response to the famous report 'Our Common Future', ESCAP was requested to gather relevant data and produce a report called 'The Environment Situation in the Asia Region in 1990'. In 1990 was held the first Meeting of ESCAP Environment Ministers, where that report was submitted. In 1995, the second Meeting of ESCAP Environment Ministers was held, and the third meeting was held in 2000. The frequency of the ministerial meetings was low (every five years), and they did not have a large impact.

Japan's Environment Agency initiated the establishment of another sectoral regime named ECO-ASIA. The first ECO-ASIA Meeting of Environment Ministers was held in 1991. Since then, the Meeting of Environment Ministers or Senior Officials has basically been held annually. Similar to ESCAP, ECO-ASIA also engages in basic information-gathering and attempts at forecasting.

Role of the comprehensive regime: APEC   In the Asia-Pacific region, there was no intergovernmental comprehensive regime comparable to the EC/EU in the European region until the end of 1980s. There were attempts to establish intergovernmental comprehensive regimes, but because of resistance mainly from the Southeast Asian countries, those attempts failed. As a result, only NGOs (non-governmental organizations) have been working until recently. In 1967, based on the initiative of the Japan–Australia Economic Commission, PBEC (Pacific Basin Economic Commission), a network of business communities, was established. In 1968, PAFTAD (Pacific Area Free Trade and Development Conference), a network of academic scholars (mainly economists), was established through the initiative of Professor Kojima. PBEC and PAFTAD have been sources of informal policy networks at the Asia-Pacific level since then. In the late 1970s, there were several attempts to establish an intergovernmental comprehensive regime (e.g., the Pacific Economic Community proposal by PBEC, the Organization for Trade and Development proposal by H. Patrick and P. Drysdall, etc.), which in the end never came into being. However, based on cooperation between Japan and Australia, PECC (Pacific Economic Cooperation Council) was established in 1980. PECC was a specific kind of NGO, composed of the representatives of government, business and academia from each country. Finally, based on an Australian initiative and supported mainly by Japan's MITI (Ministry of International Trade and Industry) and Korea, APEC was established in 1989 as the first intergovernmental comprehensive regime at the Asia-Pacific level (Shiroyama 2000b).

Initially, APEC's major field of interest was general economic cooperation. However, since 1993, mainly based on US pressure, the heart of the APEC agenda moved to a region-wide, liberal trade and investment regime. The heads of states agreed in Bogor, Indonesia, in November 1994 to reduce trade and investment barriers by 2010 for the developed countries and by 2020 for the developing countries (Kikuchi, 1995).

This trade liberalization pressure produced reactionary opinions that argued that regional economic cooperation necessitated the creation of a regional framework for environmental governance. In practice, in March 1994, the first APEC Meeting of the Environment Ministers was convened in Canada based on the initiative of Canada. It adopted two documents. The first one, called the APEC Environment Vision Statement, said that environmental protection could not be separated from economic growth and that every working group of APEC should put environmental considerations into each programme. The second one, called the Principal Framework to Integrate Economy and Environment in APEC, provided several principles concerning sustainable development, internalization of external costs, science and research, technology transfer, preventive approach, trade and development, and environmental education and information, etc. Also, at the APEC Osaka Summit in 1995, an emphasis on long-term problems such as population, food, energy and environment was incorporated into the APEC Action Declaration of Economic Leaders. The second APEC Meeting of Environment Ministers was held in 1996 in the Philippines (Zarsky and Hunter, 1997).

Three points should be made clear about APEC's environmental activities.

The first point concerns the organizational arrangement of APEC in environmental policy. At the first APEC Meeting of Environment Ministers in 1994, Canada proposed to establish an independent working group for environmental policy. However, because of the opposition from other countries, APEC did not establish an independent working group for environmental policy. Instead, the APEC Vision Statement stipulated that every working group should put environmental considerations into each programme. In one sense, it can be said that APEC was employing the advanced integration approach that the EC has emphasized recently in its environmental policy. In another sense, it can be said that the integration approach at an early stage somewhat weakens the position of environmental policy in the overall policy mix.

The second point concerns the relations between APEC and the ad hoc regimes. The interaction between APEC and the ad hoc regimes has been very weak compared to the interaction between the EC/EU and the ad hoc regimes. In East Asia, the ad hoc regimes are rather responsive to global-level events, such as at UNCED, etc. Regional sea regimes were established under the auspices of UNEP, and ESCAP was responsive to the report of the Bruntlant Commission. ASEAN also changed its ASEP3 to an Action Plan responding to Agenda 21.

The third point is the role of the consensus-oriented and discretionary mechanism in the APEC environmental regime. The main instruments of the APEC environmental policy are information exchange, technology transfer and technical cooperation. Legal instruments are not introduced in practice. The topic of trade and environment has been put on the agenda, but APEC has not produced any specific legal instruments yet. Instead, it can be said that the consensus-oriented and discretionary mechanism are employed in the APEC environmental regime.

Comparative characteristics of environmental cooperation in East Asia

As the historical developments above show, in Europe the development of the ad hoc regimes preceded the effective development of the comprehensive regime. The ad hoc regimes such as regional sea regimes and the acid rain regime were initiated by Scandinavian countries. These ad hoc regimes were not only established effectively before the comprehensive environmental regime but also played important roles in the actual development of the comprehensive environmental regime. In that sense, it is possible to say that the environmental regime in the European region has been built based on the bottom-up approach. In addition, the legal mechanism performs important roles in the EC environmental regime. In resolving the conflict concerning the interpretations about these legal instruments, the ECJ has played an important role.

Contrary to the experience in the European region, the ad hoc regimes in East Asia were peripheral phenomena and did not play important roles in the actual development of the comprehensive environmental regime in East Asia. Recently APEC, as a comprehensive environmental regime, has been engaged in environmental programmes, but APEC is acting independently and the ad hoc regimes do not have enough interaction with it. In that sense, it is possible to say that the environmental regime in East Asia is built based on the top-down approach. In addition, neither ad hoc regimes nor APEC used legal instruments, but rather relied on weak measures such as planning and programmes.

2.   New initiatives for environmental cooperation in East Asia since the mid-1990s

In the East Asian region, ASEAN and the South Pacific region historically had rather active subregional environmental regimes. However, the core Northeast Asian region (Japan, Korea, China, etc.) lacked subregional regimes even though the objective need for them seemed to be very large. This seems to be a reflection of the characteristic feature of the East Asian environmental cooperation, that is, weak ad hoc regimes. However, recently there has emerged a new tendency to establish subregional environmental regimes in this Northeast Asian region (Shiroyama, 2001).

First, the Meeting of Senior Officials on Environmental Cooperation in Northeast Asia was held in 1993 in Seoul, and attended by Japan, China, Korea, North Korea, Mongolia and Russia. At this meeting, the NEASPEC (North East Asian Subregional Programme of Environmental Cooperation) was launched.[2] This meeting was initiated by the foreign ministries of each country mainly as a follow-up to the 'Regional Strategy on Environmentally Sound and Sustainable Development in Asia and the Pacific' (ESCAP in 1991) and Agenda 21 at UNCED in 1992. The Senior Officials Meetings have been held several times since then. The sixth meeting in Seoul in 2000 adopted the Vision Statement for Environmental Cooperation in Northeast Asia and recommended that the members should effectively implement its components so as to make a valuable contribution to the environmental enhancement in the region. Following the leadership of Korea, the Meeting of Senior Officials requested the ESCAP secretariat to administer a 'core fund' for this cooperation, and Korea contributed US$100,000 to this fund. Also under this framework of NEASPEC, NEACEDT (North East Asia Centre for Environmental Data and Training) was established in the Ministry of Environment of Korea as a clearing-house of information for the purpose of environmental monitoring, data analysis and capacity building.

Second, in the Northwest Pacific region, countries have been discussing the regional sea programme since 1989. At the first intergovernmental meeting in Seoul in 1994, Japan, China, Korea and Russia adopted the 'Action Plan for the Protection, Management and Development of the Marine and Coastal Environment of the Northwest Pacific Region' (NOWPAP).[3] At the intergovernmental meeting of 1999, the Data and Information Network Regional Activity Centre in China, the Special Monitoring and Coastal Environmental Assessment Regional Assessment Activity Centre in Japan, the Marine Environmental Emergency Preparedness and Response Regional Activity Centre in Korea and the Pollution Monitoring Regional Activity Centre in Russia were designated as the Regional Activity Centres of NOWPAP. The sixth intergovernmental meeting in 2000 in Tokyo established a new project called 'Assessment and Management of Land-based Activities'.

Third, EANET (Acid Deposition Monitoring Network in East Asia) was established, following the leadership of Japan, to (1) carry out monitoring of acid rain deposition by harmonized methodologies in East Asia and (2) create a common understanding of the state of acid rain deposition to provide the scientific basis for further steps such as measures to reduce adverse impacts on the environment caused by acid deposition.[4] In 1998, EANET started its preparatory phase activities and the Interim Network Centre (INC) was established in Niigata, Japan. The members of EANET were China, Indonesia, Japan, Malaysia, Mongolia, the Philippines, North Korea, Russia, Thailand and Vietnam. Concerning funding, the principle of internal funding for national monitoring activities has been largely successful. INC provided only limited assistance to national monitoring. On the other hand, EANET has provided various chances for training of related experts. So far, EANET has been focusing on monitoring, and its impact on emissions is not yet clear, especially its impact on emissions from each country.

The above-mentioned development of the subregional and sectoral environmental regimes in the Northeast Asian region over the last several years is remarkable, considering the historical absence of subregional and sectoral environmental regimes in this region. Their activities are still fragmented and are not well coordinated; however, we can easily understand the importance of interaction among various ad hoc regimes at the field level if we recall the historical development in the European region.

In addition to the emergence of subregional and sectoral environmental regimes in the North Asian region, we are witnessing the emergence of environmental cooperation at the ministerial level in this region.[5]

Following mainly the leadership of the Korean government, the first Tripartite Environmental Ministers Meeting (TEMM) attended by China, Japan and Korea was held in Seoul in 1999. The ministers recognized that China, Japan and Korea were playing important roles in the economic and environmental cooperation in the Northeast Asian region and also that close cooperation among the three countries was indispensable to sustainable development in Northeast Asia. Raising awareness, activating information exchange and strengthening cooperation in the area of environmental research relating to environmental technology and industry were recognized as the concrete agenda.

It was said that Korea initially tried to establish legal instruments to have an impact on the so-called pollution producing-activities in China. However, members, including Korea, came back to the original position of relying on consensus-oriented measures rather than pushing for a legalized solution.

In 2000, the second TEMM was held in Beijing. TEMM noted that it was necessary to promote exchanges and cooperation in various forms among central governments, local governments, science and research institutes, enterprises and NGOs. Ministers also expressed their wish to have more concrete project cooperation, in particular projects on raising consciousness concerning the environmental community, freshwater (lake) pollution, land-based marine pollution prevention and cooperation in the field of environmental industry. Ministers also recognized the importance of activities relating to EANET and NOWPAP.

In 2001, the third TEMM was held in Tokyo. Ministers exchanged views on recent progress in environmental management in their countries and noted the high recognition given to the progress of TEMM by the Trilateral Meeting of the Leaders of Japan, China and Korea in November 2000. Concerning concrete issues, ministers expressed great concerns about the degradation of the natural environment in northwest China and shared the recognition that the three countries should promote cooperation for systematic studies on sand dust (yellow dust or kosa), which is exacerbated by the soil degradation.

Annual meetings were also held in 2002–4. Concrete projects under way under TEMM are: joint environmental training, development of an environmental education network, official website creation and maintenance, freshwater (lakes) pollution prevention, collaboration in environmental industry development and ecological conservation in northwest China. The dust and sandstorm (DSS) issue has been sensitive among the three countries because Korea and Japan think they are victim countries in the 'downstream', while China insists China is, too, a victim country. At the sixth meeting held in Tokyo in December 2004, the 'first four' nation ministers' meeting was held, and included among the invitees Mongolia's Minister of Natural Environment and some representatives of international organizations. It is worth noting whether this meeting has been able to promote effective regional cooperation on DSS in Northeast Asia.

Also, at the sixth meeting, the building of an 'eco-recycling economy' to promote the 3Rs (reduce, reuse and recycle) was focused upon as hot issue and addressed in a joint communiqué to support China's initiative to incorporate this idea into the next (11th) national Five-year Plan (2006–10). For the sake of promoting cooperation among environmental NGOs in the three countries, in late 2000 Japan's Ministry of Environment granted additional financial support for the Japan Fund for Global Environment (JFGE), which was established in 1993 to support environmental conservation activities by NGOs.[6] Since then, the three countries' environmental cooperation has been one of the priority targets in JFGE.

Those developments at the ministerial level are also remarkable considering the political attention given to environmental issues in the Northeast Asian region and the potential interaction between the regime at the ministerial level and subregional/sectoral regimes (the second TEMM paid attention to the activities of EANET and NOWPAP).

In recent events, it seems that Korea is a major initiator for the multilateral environmental cooperation regime in Northeast Asia. Japan played a major role in establishing EANET, but was relatively reactive in other cases.

Above all, to design an effective regional environmental regime, coordination among various components is very important. First, there has to be coordination among various initiatives in each sector, such as regional sea and acid rain. Second, the effective interaction between ad hoc regimes and comprehensive/high-level regimes (such as APEC and TEMM) has to be structured. Third, the international cooperation programmes have to be integrated into the regulatory regimes to cope with the problems arising from the difference in economic development levels (while paying attention also to the incentives of private enterprises).

3. Forms and effectiveness of environmental aid: focusing on the case of Japan's environmental aid to China

Environmental aid as embedded modality of environmental cooperation

One of the important characteristics of the environmental cooperation regime in the East Asian region is that the regional environmental policy is also embedded in the international aid programmes. So it can be said that one aspect of international aid is the invisible environmental regime. These aid programmes are closely related to regime setting; interaction between various actors relating to such aid highlights the important aspects of operation of the regional environmental regime. Moreover, the environmental policy is embedded not only in the technical cooperation component of the multilateral programmes but also in many other bilateral aid programmes, such as Japanese ODA and others.

Functionally, it is necessary to incorporate local considerations, including considerations relating to stages of economic development, into the international framework. For example, since China, which is a main source of emissions in East Asia, is one of the developing countries, international aid is necessary at least transitionally. Even though environmental technology transfer can be realized through the market mechanism and internal conditions are important, the necessity for international support will not vanish.

Japan has been a major donor for environmental aid to China. Japan's environmental aid to China increased especially after the late 1990s because concerns increased among Japanese people over the impact on Japan of emissions from China and because it is relatively easy to justify aid to China, as it industrializes, for social issues such as environment and poverty.

Here, the forms and effectiveness of environmental aid will be analysed as a type of modality of environmental cooperation, mainly using the case of Japanese environmental aid to China.

Types of environmental aid

Since the early 1990s, increased attention has been focused on the environmental impact of coal combustion in China as a cause of local, regional and global environmental problems. In China, coal still constitutes more than 70 per cent of the total primary energy source, and there seems little prospect of a large-scale conversion to oil, natural gas, or a renewable energy in the immediate future.

To address the environmental concerns related to the high level of coal use in China, there are various attempts to improve the technologies for coal combustion used in China. For example, Japan, together with Germany and the World Bank, have drastically increased both direct and indirect assistance to China in this area. For its part, China is also in the process of introducing a new regulatory framework to control emissions, which increases the demand for environmental technologies. Here, international environmental aid programmes relating to improvement of coal combustion are presented as concrete cases.

There are basically three types of international cooperation programmes (Shiroyama, 1998). The first type is composed of project finance plans for coal-fired power plants developed by the Japan Bank for International Cooperation (JBIC), together with the World Bank, the Asian Development Bank and other institutions. These programmes provide large-scale, more efficient boilers for coal-fired power plants. As the number of power plants is limited and because foreign companies can receive contracts for this work, international donors often implement this approach.

The second type directly involves technology transfer projects. These projects try to facilitate low-cost domestic production of environmental facilities in order to assist development of the supply-side capacity. Donors have an interest in facilitating technology transfer because it can reduce total project costs.

Green Aid Plan projects of Japan's Ministry of Economy, Trade and Industry support the proliferation of simplified flue gas desulphurization (FGD) and other technology that is suitable in the Chinese context. These projects also try to increase the localization of production to reduce costs. Similarly, the Industrial Boiler Project of the Global Environmental Facility (GEF) implemented by the World Bank tries to promote the production and diffusion of high-efficiency industrial boilers by supporting the cost of licensing by foreign companies to nine Chinese local manufacturers (GEF, 1996).

The third type of programme involves institution-building and model city projects. These projects try to stimulate the domestic demand for environmentally friendly facilities in China. Institution-building projects provide incentives for local enterprises by strengthening local environmental regulations, such as emission fees. There is one JBIC project in Liuzhou in the Guangxi autonomous region, and the Japanese government also has three more large-scale environmental model city projects in Dalian in Liaoning Province, Guiyang in Guizhou Province and Chongqing in Sichuan Province. Model city projects allow local people to gain first-hand experience in developing environmentally friendly facilities. Similarly, the World Bank's Chongqing Industrial Reform and Pollution Control Project in Sichuan Province provides an industrial reform fund (for environmental improvement and other purposes) on the condition that the Chongqing City Government will introduce an SO_2 emission fee system (World Bank, 1996).

In addition to these programmes specifically focused on coal combustion, there is general technical assistance relating to the environment. Concerning general technical assistance relating to the environment, many interesting initiatives were undertaken, one of which was the establishment of the Sino-Japan Friendship Centre for Environmental Protection (SJC) (Otsuka, 2004). The SJC was built by Japanese grants-in-aid in 1996, upon agreement between Japanese and Chinese governments in commemoration of the tenth anniversary of the Japan–China Peace Friendship Treaty. The centre belongs to China's State Environmental Protection Administration (SEPA). The centre consists of both hardware and software. The hardware of the centre consists of offices, facilities and environmental equipment. Its software includes Japanese experts on environmental analysis and policy who are working with Chinese staff from several institutes of SEPA to develop cooperative action plans between the two countries and to give technical advice to SEPA.

SJC works in cooperation with the Japanese International Cooperation Authority (JICA) on technical cooperation projects, which have been conducted over the course of ten years in three phases. In the first phase of the SJC project in 1992–5, a personnel training programme, in which Chinese staff visited Japan, was conducted. In the second phase (1996–2001) and its follow-up stage (2001–2), when SJC was officially launched as branch institute of SEPA, there were initiated some concrete cooperative projects, such as environmental monitoring, pollution control technology research, environmental policy study, personnel training and environmental information development. During the first half of the third phase (2002–4), for the further development of the results of the second-phase projects, there were four focal projects, namely those(1) addressing the air pollution problems that extend to vast areas of China (sand storm, acid rain and particulates); (2) seeking measures for improving environmental management standards (pollution control manager system and training for local environmental protection bureau chiefs); (3) dealing with chemical substances that have become new threats; and (4) seeking measures for environmental protection in the region subject to the Great Western Development Project.

In the latter half of the third phase (2004–6),[7] there are two types of projects: (1) policy-institution assistance for promotion of an eco-recycling economy, promotion of enterprises' environmental protection manager systems, constitution of basic environmental law, implementation of environmental impact assessments and environmental model city planning; and (2) technology transfer assistance for dioxin analysis technology, POPs analysis technology, analysis of urban air contamination particulates' sources, including sand storms, and promotion of solid waste recycling. The general tasks for cooperation include capacity-building for monitoring acid rain and assistance for training for local environmental protection bureau chiefs. In addition to these projects, SJC also coordinates with other JICA schemes to conduct training courses for pollution control managers and feasibility studies for technical assistance. Since March 2003, an expert from JBIC has been sent to SJC to coordinate yen loan projects by JBIC in conjunction with technical corporation projects by JICA in the field of environmental protection.

Implementation of environmental aid

Implementation of environmental aid needs to be analysed to understand the dynamism of the interaction between international and domestic levels. Two examples of environmental aid to China will be analysed, to give concrete examples. One is the Green Aid Plan by Japan's Ministry of Economy, Trade and Industry. The other is the Model City Project by JBIC.

In the framework of the Green Aid Plan (GAP) by the Ministry of Economy, Trade and Industry, various demonstration experiments have been conducted. The scale of GAP projects is about 2–3 billion yen per project. It was expected that a simplified version of FGD would be developed through a demonstration project by the Japanese side and the Chinese side. However, because of a request from China, which asks for perfect quality under the present circumstances, the cost is high since the ratio of the parts supplied from Japan is high (and it is said that the cost would drop to one third if most of the equipment were produced domestically, for example in the case of the simple desulphurization facility at Zhangshou chemical factory in the Chongqing suburbs).[8] Since the Chinese parties that are involved are users, not producers, of FGD equipment (i.e. the former Bureau of Electric Power Industry and the former Bureau of Chemical Industry, etc.), there was a difficulty in pursuing lower cost.

For the facility above, the annual operation cost of the FGD equipment is about 600,000 yuan; meanwhile, the annual amount of saving of pollution discharge fee is only 150,000 yuan. This means that the running cost is four times larger than the amount saved on pollution discharge fees. Moreover, if the capital amortization is included, even in the case when the facilities are produced locally, the annual cost will be 2 million yuan. Even though it is true that there are prospective annual sales of 185,000 yuan-worth of gypsum, which is a by-product, it is not sufficient to justify the operating cost.[9] It is generally said that it is necessary to set the pollution discharge fee five times higher in order for a company to have an incentive. (Some local governments do try to experimentally increase the pollution discharge fee by several times.)

In addition, disputes over licences have occurred. GAP does not include royalty cost at the time of technology diffusion, and it was vague on which part of the results of R&D developed jointly by demonstration projects belong to China. More fundamentally, although there was an incentive on donor side to conduct the demonstration experiment in China as part of a marketing strategy, there was difficulty in technology transfer following the demonstration stage, due to issues involving the security of intellectual property rights in China and issues of fear of a boomerang effect, etc. (Shiroyama, 2000a, p. 87).

Based on project experiences in Liuzhou, in the Guangxi autonomous region, by JBIC, the Japanese government and JBIC established three more large-scale environmental model city projects in Dalian, in Liaoning Province, Guiyang, in Guizhou Province, and Chongqing, in Sichuan Province. The budget of model city projects was allocated in 1999 and 2000. The rationale of the model city projects is to stimulate the domestic demand for environmentally friendly facilities in China. It is expected that institution-building components will provide incentives for local enterprises for behavioural change and for capital investment for facilities upgrades by strengthening local environmental regulations involving monitoring, emission fees and suspension of operation. However, the effectiveness of institution-building and regulation is different depending on the attitudes of local government authorities. Even if the central government (especially SEPA) is positive about strengthening environmental regulations, local governments play major roles in the implementation. It is said that local government is very positive about strengthening environmental regulations in the case of Liuzhou and Guiyang (in those cases, training of local government officials for environmental protection through SJC was effective), but that local government is not so positive in the case of Chongquing.[10]

There seems to be no other city following the model set by the three model cities. Even though the rationale of the model city projects is to stimulate the domestic demand for environmentally friendly facilities in China, this has not materialized yet. The main reason seems to be the inadequacy of the financial resources allocated to local governments for environmental investments. In other words, models have to adapt to local conditions, including economic conditions.

Factors determining the effectiveness of environmental aid

There are several cross-cutting factors that influence the nature and implementation of international aid (Shiroyama, 1998, 2000a).

The first factor is domestic environmental regulation. In general, enterprises will rarely participate in emission control projects without the threat of penalties. In China, an 'Acid Rain Control Area' and an 'SO_2 Control Area' were set up in 1998 to control the SO_2 emissions. In those areas, FGDs are required for future power plants that will use coal with a sulphur content exceeding 1 per cent. Additionally, an SO_2 emission fee of 0.2 yuan per 1 kg of SO_2 is levied on every source of SO_2 emission. Even with this measure, many still believe that the emission fee is too low to induce enterprises to invest in the FGD.

Considering the conditions above, it is necessary to increase the emission fee. On the other hand, since the pollution discharge fee serves, in many cases, as the main source of revenue for the local environmental protection bureaus, a perverted incentive structure has arisen. That is, if we think of pollution discharge fees as an economic measure for environmental conservation, it is necessary for the emission fee to exceed the installation expense of pollution control equipment. However, since the local environmental protection bureau will lose its main source of revenue if it sets the fee high enough to cause companies to install pollution control equipment, it has an incentive to set the level of the pollution discharge fee below the level of the installation expense for pollution control equipment. As a result, the discharge by companies will continue.

The second factor is the financial environment of the user of technologies. In general, enterprises will voluntary participate in energy efficiency improvement projects if the energy cost savings that result from increased efficiency is larger than the investment cost for efficiency improvement. However, this is not what is occurring in China, and it is because of the financial difficulties in that country (Lardy, 1998). At present, we are witnessing an 'energy paradox' in China, i.e., the phenomenon that even profitable projects (such as energy efficiency improvement projects in which the energy cost saving through efficiency improvement is greater than the investment cost for efficiency improvement) are not being undertaken (Cui, 1998). There is a clear shortage of financial support for potentially profitable projects, a situation that has been exacerbated by the recent domestic debt crisis in China.

The third factor involves issues of organization and agency. Donor agencies have to work with their counterparts in China to implement international aid programmes. The choice of recipient agencies will have a substantial effect on the nature of future projects. In the case of Japan's Green Aid Plan, Japanese agencies have been working mainly with Chinese vertically fragmented sectoral line agencies, such as the former Ministry of Electric Power and the former Ministry of Chemical Industry. As those agencies are user agencies, they have a greater interest in obtaining improved technology than in developing manufacturing capabilities and reducing costs. This is one of the reasons why Japan has had difficulty in persuading the Chinese side to promote localization. On the other hand, in the case of the GEF (Global Environmental Facility) Industrial Boiler Project, GEF has been working with the former Ministry of Machinery, which is focused on the supply side. This is one of the reasons why the GEF project can deal with the localization issue directly. Additionally, a multiplicity of jurisdictions sometimes complicates the process even further. This kind of fragmented structure is prevalent not only in the government but also in industrial organization in China. For example, in China there is no concept of an 'engineering company' where various component parts are integrated in a consistent manner.[11] So it is very hard to localize overall manufacturing capabilities even when the local party can produce components.

In addition to those domestic factors mentioned above, there are other factors at the international level.

The first of these are rules related to trade. The OECD has long regulated export credit to avoid distortions in competition. Based on that experience, in 1991 OECD countries agreed to establish a series of rules (the Helsinki Package) to restrict 'tied aid' (aid that has to be procured in the donor country).[12] Under that package, 'tied aid' is prohibited if the Concessionary Level (CL) is less than 35 per cent. If the CL is between 35 per cent and 80 per cent, 'tied aid' is permitted only when the project is deemed 'commercially nonviable' (CNV) and prohibited if judged to be 'commercially Viable' (CV). The decision about whether a project is CV or CNV is made on a case-by-case basis, according to the calculation of cash flow in the abstract market. While many projects in the railway and hydropower sectors are approved as CNV, projects related to coal-fired power plants and industrial boilers are usually not. The Helsinki Package may affect the provision of aid in two ways. First, because a reduction of 'tied aid' may weaken domestic support for international aid activities, the total amount of aid may be reduced. Second, allocation of aid may be biased against CV projects, such as clean coal technology.

Second is the attitude of NGOs, even though this factor is not intense in Japan. Some NGOs strongly oppose aid to fossil fuel projects like coal-fired power plants. For example, the Washington, DC-based Sustainable Energy and Economy Network (SEEN) has criticized the World Bank for supporting fossil fuel projects, a source of global warming, despite the Bank's acknowledgement that climate change is disastrous for poor nations.[13] Even though coal is a major source of energy in China and there is room for utilization efficiency improvement through modernization of coal combustion facilities, SEEN is against the idea of aid agencies or export–import banks supporting fossil fuel projects in developing countries such as China.[14] Although the World Bank is sensitive to this criticism, it continues to support coal-fired power plant projects. In case of GEF, however, NGO objections and the political atmosphere in Washington make it very difficult to support coal-related projects. Environmentalist NGOs and trade advocates can make a tactical alliance against the 'tied aid' projects on coal utilization efficiency improvement.

Third, what is more, there is also the question of the incentive of foreign companies. There is a potential gap between the interests of donors that promote technology transfer to reduce project costs and the interests of private enterprises, which like to maximize their benefits. In the case of Japan's Green Aid Plan, manufacturers in Japan had few incentives to promote localization but have been motivated to participate in demonstration projects knowing that the experience gained will be useful when they develop plans to export their facilities.[15] In the case of GEF's Industrial Boiler Project, the World Bank tried to identify foreign companies to provide licences for small- and medium-size industrial boilers to be manufactured. However, large companies such as ABB and Ebara declined to participate because the contract was too small and involved complicated procedures. At present, only small companies from North America and Europe are participating.[16]

Also, there are many potential conflicts relating to intellectual property rights between foreign companies and recipients, as mentioned before. In the case of Japan's Green Aid Plan, there are reports of disputes concerning the distribution of rights arising from the joint demonstration process. In addition, there is the problem of legislation concerning licences in China. The duration of licences for the technology concerned was to be limited by legislation to a shorter period of time than the licence period usually seen in other countries (International Centre for Environmental Technology Transfer, 1996, p. 36). As a result, the incentive of technology owners was diminished. One solution to this is for foreign companies to establish JVs (joint ventures) with Chinese companies and to license the JVs. In this case, after the end of licence periods, the foreign companies are able to enjoy profits because foreign companies retain their positions as stockholders of JVs. However, since this method also has the disadvantage of requiring a large initial investment, it seems that there are few companies that choose this way.[17]

The above factors have to be incorporated when designing effective international cooperation that is needed in regions such as East Asia, which is comprised of both developed and developing countries.

Inherent limits of Japan's environmental aid to China

There are two difficulties in Japan's policy for environmental aid to China.

The first one relates to the justification of environmental aid to China. Japan has been a major donor of environmental aid to China, especially since the late 1990s. Japan increased the ratio of its environmental aid because of increased concerns on the part of Japanese people over the impact on Japan of emissions from China and because it is relatively easy to justify aid to China, as it industrializes, for environmental protection that mitigates damage to Japan. For monitoring the impact of Chinese emissions on Japan, Japan has been a leader in the establishment and operation of EANET.

However, despite the Japanese government's justification for environmental aid to China, it is not easy to assess the impact of cross-border pollution.[18] Even with the advances in monitoring and data-sharing on regional environmental quality, as with EANET, it is not easy for scientific assessment to detect a clear impact from cross-border pollution, such as acid rain. In 2002, Japan's Ministry of Environment released a survey on acid rain control, which was conducted not only by monitoring acid deposition in soil, vegetation and inland water, but also by using other research, including research from EANET (Ministry of Environment, 2002). It revealed that it is difficult to evaluate the impact of acid deposition on the natural ecosystem in the Japanese islands.

The second difficulty relates to the model that Japan has been using for environmental aid. Some are of the opinion that Japanese experiences with pollution control can be transferred to China wholesale. Although it is true that there is much useful experience in Japan, there were also certain unique features in Japan's pollution experience (e.g., corporate reorganization and economic reform were performed in the 1950s in Japan, before pollution control became required in the 1960s). It seems that it is not appropriate to emphasize 'Japan's experience', without recognizing the structural differences between the previous conditions in Japan and the current conditions in China (where economic reform and environmental measures are being called for simultaneously). Although there is a simplistic tendency to introduce FGD, which was a key point in Japan's conquest of its air pollution problem, it is necessary to search for technical adaptations that are suited to the current situation of China in programmes such as GAP.

It is also necessary to adapt to the imperfect nature of implementation of environmental regulations in China. However, there is also some progress in response to social pressure, such as the improvement of monitoring systems and equipment. This point is also supported by 'Japan's experience' with pollution control. Because of the positive effect of social pressure, the organization of an environmental movement by non-governmental organizations in China is indispensable. An important question in socialist states such as China is how an environmental movement can be differentiated from a political opposition movement against the government.

4. Conclusion

In this chapter, the characteristics of the regional environmental cooperation regime of the East Asian region are analysed in comparison with the regional environmental cooperation regime of Europe. In contrast to Europe where ad hoc regimes were not only established effectively before the comprehensive environmental regime but also played important roles in the actual development of the comprehensive environmental regime, the ad hoc regimes in East Asia were peripheral phenomena and did not play important roles in the actual development of the comprehensive environmental regime in East Asia. The APEC regime has been engaged in environmental programmes as a comprehensive regime, but APEC is acting independently and the ad hoc regimes do not interact with it. In particular, the Northeast Asian region (Japan, Korea, China, etc.) lacked subregional regimes even though the objective need for them seemed to be large.

Recently, however, there has emerged a new tendency to establish a subregional environmental regimes in the Northeast Asian region, such as the ad hoc regimes for regional sea and acid rain (NOWPAP and EANET) and TEMM between Korea, China and Japan. Although those tendencies are interesting, how those developments will be institutionalized is not clear yet.

In addition, the forms and effectiveness of environmental aid were analysed, using mainly cases of Japan's environmental aid to China. Factors such as domestic environmental regulations, financial environment, organizational coordination, international rules and roles of NGOs and foreign companies were identified. As international aid and cooperation programmes are sometimes components of and closely related to the programmes of regional environmental regimes, those factors that were identified (such as institutionalization of domestic regulation, organizational coordination and compatibility with international systems) are also important for the effectiveness of regional environmental governance.

Finally, I would like to emphasize the issue of Japan's strategy for regional environmental cooperation. Japan directly or indirectly tries to promote regional environmental adaptation through the use of environmental aid. However, the effectiveness of those activities seems to be limited by the model Japan has been employing (e.g., simple strict enforcement and expensive technology). Although it is true that there are many useful experiences in the Japanese model of environmental adaptation, there were certain unique conditions in the case of Japan's pollution control. It seems that it is not appropriate to emphasize 'Japan's experience', without recognizing the structural differences (e.g., the sequence and stage of economic development and the existence of a social movement) between the previous condition of Japan and the local conditions where aid programmes are being targeted. In addition, in recent years, there have developed nascent discussions over when to cut off Japan's official development aid to China. This trend in discussions is due to a number of factors, such as China's growing economy, China's military threat, Japan's economic crisis and growing antipathies among people in both countries, which further complicate the situation.

Notes

1. Refer to Shiroyama (1997) and Shiroyama (2001) concerning comparisons of the historical structure of environmental regimes in Europe and East Asia.
2. http://www.temm.org/docs/coop/mulcoop_view.html?seq=00004&menu=6
3. http://www.temm.org/docs/coop/mulcoop_view.html?seq=00002&menu=6
4. http://www.eanet.cc/jpn/
5. http://www.temm.org/docs/main.html
6. http://www.erca.go.jp/jfge/english/index.html
7. http://www.zhb.gov.cn/japan/e_index.htm
8. An interview at Chongqing Zhangshou Chemical factory in July 1997.
9. An interview at Chongqing Zhangshou Chemical factory in July 1997.
10. An interview at JBIC, in February 2005.
11. An interview at Ebara in November 1999.
12. Refer to Japan Consulting Association (1998) for a general view of the Helsinki Rule and to Owen (1998) for its implementation.
13. Concerning activity of SEEN, see its website at http://www.seen.org
14. This is based on the suggestion of Dr Peter Evans (Massachusetts Institute of Technology).
15. This is based on interviews at related companies during October and November 1999.
16. An interview at the World Bank in October 1997.
17. Interviews at Toshiba and Hitachi in November 1999.
18. Concerning these findings, the author was given valuable suggestions by Professor Jusen Asuka, Tohoku University.

References

ASEAN Secretariat (1995) *Trade and Environment: Issues and Opportunities*, Jakarta, ASEAN.

Cui, Zhiyuan (1998) 'The Incentive and "Energy Paradox" in the Chinese Coal and Industrial Boiler Industry', Cambridge, MIT Centre for International Studies.

GEF (1996) *China: Efficient Industrial Boilers (Report No. 16132-CHA)*, Washington, DC, World Bank.

Haas, P. M. (1993) 'Protecting the Baltic and North Seas', in P. M. Haas, R. O. Keohane and M. A. Levy (eds) *Institutions for Earth*, Cambridge, MIT Press.

International Research Centre for Environmental Technology Transfer (1996) '*Chugoku oyobi Asean sankakoku niokeru Kankyo Gijutu Itenseika no teityaku hukyu chosa*' (*Establishment and Diffusion of Environmental Technology Transfer in China and Three ASEAN-countries*) (in Japanese), Yokkaichi, International Research Centre for Environmental Technology Transfer.

Japan Consulting Association (1998) 'Surveillance Study on Possibility of the Tide Assistance Based on the International Rules' (in Japanese), Japan Consulting Association.

Johnson, S. P. and G. Corcelle (1989) *The Environmental Policy of the European Communities*, London/Boston, Graham and Trotman.

Kikuchi, Tsutome (1995) *APEC: ajiataiheiyo shinchitsujo no mosaku* (*APEC: Searching for Regional Order in Asia-Pacific*). Tokyo, The Japan Institute of International Affairs, (in Japanese).

Lardy, Nicholas R. (1998) *China's Unfinished Economic Revolution*, Washington, DC, Brookings Institution Press.

Levy, M. A. (1993) 'European Acid Rain: The Power of Tote-Board Diplomacy', in P. M. Haas, R. O. Keohane and M. A. Levy (eds), *Institutions for Earth*, Cambridge, MIT Press.

Ministry of Environment (2002). 'Dai 4 ji sanseiu taisakuchosa no torimatome nitsuite' (A Report of the fourth Survey on Acid Rain Control) Press release (in Japanese).

Otsuka, Kenji (2004) 'Nittyu yuko kankyo hozen senta' (Sino-Japan Friendship Centre for Environmental Protection') (in Japanese), in Tyugoku Kankyo Mondai Kenkyukai (study group on China's environmental problems) (eds), *Chugoku kankyo handobukku* (*A Handbook on China's Environment*). The years of 1995–6, Tokyo, Sososha, 393–4 (in Japanese).

Owen, Tony (1998) 'The Body of Experience Gained under the Helsinki Disciplines' unpublished paper.

Shiroyama, Hideaki (1997), 'Kokusai kankyo no gabanansu' ('Governance of International Environment') (in Japanese), *Sobun*, January.

Shiroyama, Hideaki (1998) 'Taityu kankyo Enjo no Kozu' ('Structural Outline of Environmental Assistance to China') (in Japanese), *Gakushikai Kaiho*, 821.

Shiroyama, Hideaki (2000a) 'Chugoku sekitan nensho mondai heno kokusaiteki taiou to Kokunai Kokusai seido no eikyo' ('International Response to China's Coal Combustion Problem, and Influence of Domestic and International Institutions') (in Japanese), *Kankyo Kyosei* (*Environmental Symbiosis*), 4.

Shiroyama, Hideaki (2000b) 'Higashiajia niokeru kokusai kihan jitugen no sosikiteki kiban' ('Organizational Basis for Implementing International Norms in East Asia') (in Japanese), in Y. Onuma (ed.), *Toa no Koso* (*Vision of East Asia*), Tokyo, Tshikuma Shobo.

Shiroyama, Hideaki (2001) 'Environmental Cooperation in the Northeast Asia? Comparison with Europe and the Concrete Agenda for the Future', paper presented at KAPA (Korean Association of Public Administration) International Conference.

World Bank (1996) *China: Chongqing Industrial Pollution Control and Reform Project* (*Report No.15387-CHA*), Washington, DC, World Bank.

Zarsky, L. and J. Hunter (1997) 'Environmental Cooperation at APEC: The First Five Years', *Journal of Environment and Development*, 6 (3), pp. 222–51.

Index